electricity
one

Hayden Electricity One–Seven Series

Harry Mileaf, Editor-in-Chief

electricity one Producing Electricity □ Atomic Theory □ Electrical Charges □ Electron Theory □ Current □ Voltage □ Magnetism □ Electromagnetism

electricity two D-C Circuits □ Direct Current □ Resistors □ Ohm's Law □ Power □ Series Circuits □ Parallel Circuits □ Series-Parallel Circuits □ Kirchhoff's Laws □ Superposition □ Thevenin's Theorem □ Norton's Theorem

electricity three A-C Circuits □ Alternating Current □ A-C Waveforms □ Resistive Circuits □ Inductors □ Inductive Circuits □ Transformers □ Capacitors □ Capacitive Circuits

electricity four LCR Circuits □ Vectors □ RL Circuits □ RC Circuits □ LC Circuits □ Series-Parallel Circuits □ Resonant Circuits □ Filters

electricity five Test Equipment □ Meter Movements □ Ammeters □ Voltmeters □ Ohmmeters □ Wattmeters □ Multimeters □ Vacuum-Tube Voltmeters

electricity six Power Sources □ Primary Cells □ Batteries □ Photo, Thermo, Solar Cells □ D-C Generators □ A-C Generators □ Motor-Generators □ Dynamotors

electricity seven Electric Motors □ D-C Motors □ A-C Motors □ Synchronous Motors □ Induction Motors □ Reluctance Motors □ Hysteresis Motors □ Repulsion Motors □ Universal Motors □ Starters □ Controllers

electricity
one

HARRY MILEAF EDITOR-IN-CHIEF

revised second edition

HAYDEN BOOKS
Macmillan Computer Publishing

© 1966 and 1976 by Hayden Books
A Division of Howard W. Sams & Co.

SECOND EDITION
FIFTEENTH PRINTING—1990

All rights reserved. No part of this book shall be reproduced, stored in a
retrieval system, or transmitted by any means, electronic, mechanical,
photocopying, recording, or otherwise, without written permission from the
publisher. No patent liability is assumed with respect to the use of the
information contained herein. While every precaution has been taken in
the preparation of this book, the publisher assumes no responsibility for
errors or omissions. Neither is any liability assumed for damages resulting
from the use of the information contained herein.

International Standard Book Number: 0-8104-5945-0
Library of Congress Catalog Card Number: 75-45504

Printed in the United States of America

preface

This volume is one of a series designed specifically to teach electricity. The series is logically organized to fit the learning process. Each volume covers a given area of knowledge, which in itself is complete, but also prepares the student for the ensuing volumes. Within each volume, the topics are taught in incremental steps and each topic treatment prepares the student for the next topic. Only *one* discrete topic or concept is examined on a page, and *each* page carries an illustration that graphically depicts the topic being covered. As a result of this treatment, neither the text nor the illustrations are relied on solely as a teaching medium for any given topic. Both are given for *every* topic, so that the illustrations not only complement but reinforce the text. In addition, to further aid the student in retaining what he has learned, the important points are summarized in text form on the illustration. This unique treatment allows the book to be used as a convenient review text. Color is used not for decorative purposes, but to accent important points and make the illustrations meaningful.

In keeping with good teaching practice, all technical terms are defined at their point of introduction so that the student can proceed with confidence. And, to facilitate matters for both the student and the teacher, key words for each topic are made conspicuous by the use of italics. Major points covered in prior topics are often reiterated in later topics for purposes of retention. This allows not only the smooth transition from topic to topic, but the reinforcement of prior knowledge just before the declining point of one's memory curve. At the end of each group of topics comprising a lesson, a summary of the facts is given, together with an appropriate set of review questions, so that the student himself can determine how well he is learning as he proceeds through the book.

Much of the credit for the development of this series belongs to various members of the excellent team of authors, editors, and technical consultants assembled by the publisher. Special acknowledgment of the contributions of the following individuals is most appropriate: Frank T. Egan, Jack Greenfield, and Warren W. Yates, principal contributors; Peter J. Zurita, Steven Barbash, Solomon Flam, and A. Victor Schwarz, of the publisher's staff; Paul J. Barotta, Director of the Union Technical Institute; Albert J. Marcarelli, Technical Director of the Connecticut School of Electronics; Howard Bierman, Editor of *Electronic Design;* E. E. Grazda, Editorial Director of *Electronic Design;* and Irving Lopatin, Editorial Director of the Hayden Book Companies.

HARRY MILEAF
Editor-in-Chief

contents

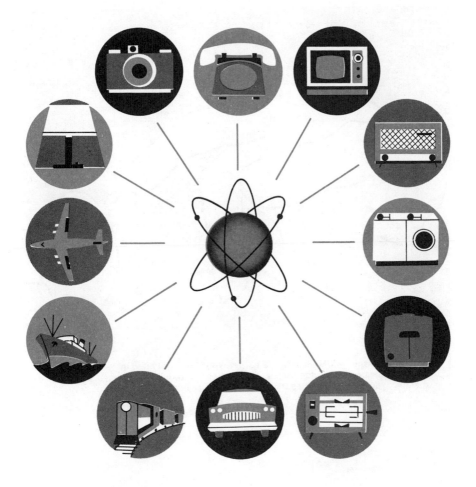

the importance of electricity

Electricity is one of the most important forms of energy used in the world today. Without it, there would be no convenient lights, no radio or television communications, no telephone service; and people would have to go without the many household appliances that are taken for granted. In addition, the field of transportation would not be as it is today without electricity. Electricity is used in all types of vehicles. When you stop and think about it, electricity is used everywhere.

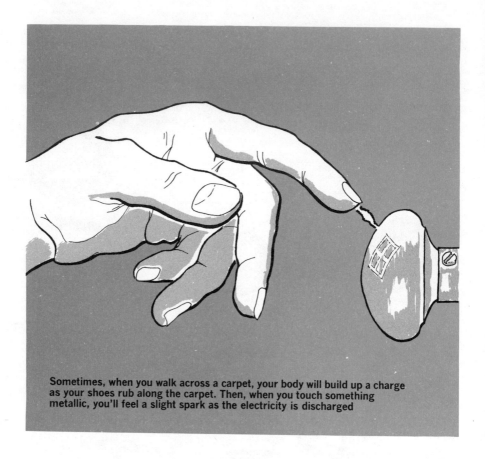

Sometimes, when you walk across a carpet, your body will build up a charge as your shoes rub along the carpet. Then, when you touch something metallic, you'll feel a slight spark as the electricity is discharged

early history

Although electricity first became useful during modern times, it was discovered 2000 years ago by the Greeks. They noticed that when a material that we now call amber was rubbed with some other materials, it became *charged* with a mysterious force. The charged amber attracted such materials as dried leaves and wood shavings. The Greeks called the amber *elektron,* which is how the word electricity came about.

Around 1600, William Gilbert classified materials that acted like amber as *electriks,* and those materials that did not, as *nonelectriks.*

In 1733, a Frenchman, Charles DuFay, noticed that a charged piece of glass *attracted* some charged objects, but *repelled* other charged objects. He concluded that there were two types of electricity.

Around the middle 1700's, Benjamin Franklin called these two kinds of electricity *positive* and *negative.*

what is electricity?

In Benjamin Franklin's time, scientists thought that electricity was a fluid that could have positive and negative charges. But today, scientists think electricity is produced by very tiny particles called *electrons* and *protons*. These particles are too small to be seen, but they exist in all materials. To understand how they exist, you must first understand the structure of all matter.

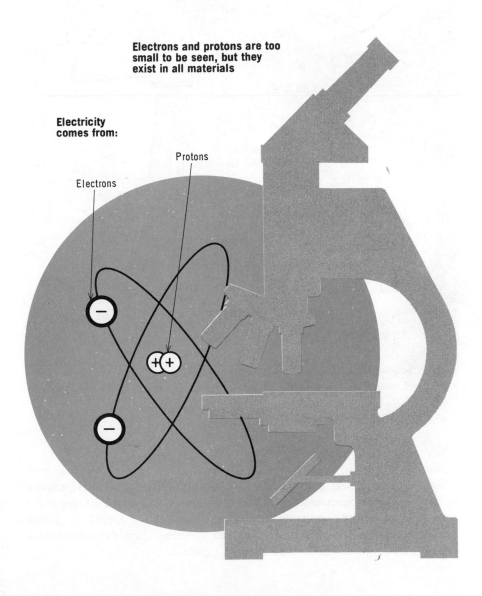

Electrons and protons are too small to be seen, but they exist in all materials

Electricity comes from:

Protons

Electrons

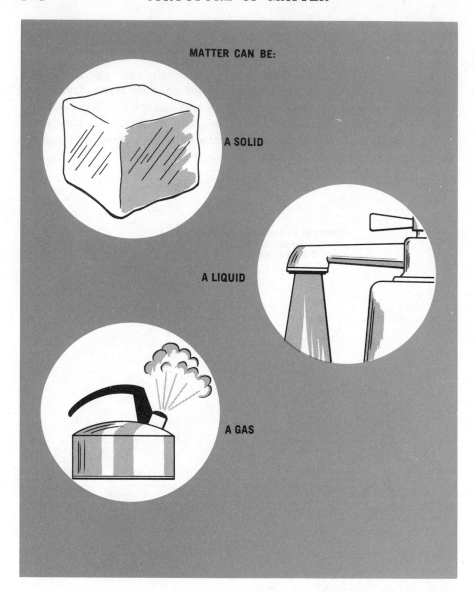

MATTER CAN BE:

A SOLID

A LIQUID

A GAS

what is matter?

Anything that you can see, feel, or use is matter. Actually, matter is anything that has *weight* and occupies *space*. It can be in the form of a solid, liquid, or gas. Stone, wood, and metal are forms of matter (solids), and so are water, alcohol, and gasoline (liquids), as well as oxygen, hydrogen, and carbon dioxide (gases).

the elements

The *elements* are the basic materials that make up all matter. Oxygen and hydrogen are elements, and so are aluminum, copper, silver, gold, and mercury. Actually, there are just a little over 100 known elements. Ninety-two of these elements are natural, and the others are man-made. In the last few years, a number of new ones were made, and it is expected that there are many more still to be produced.

Everything we see about us is made from the elements. But the elements themselves cannot be produced by the simple chemical combination or separation of other elements.

THE NATURAL ELEMENTS

Atomic Number	Name	Symbol	Atomic Number	Name	Symbol	Atomic Number	Name	Symbol
1	Hydrogen	H	32	Germanium	Ge	62	Samarium	Sm
2	Helium	He	33	Arsenic	As	63	Europium	Eu
3	Lithium	Li	34	Selenium	Se	64	Gadolinium	Gd
4	Beryllium	Be	35	Bromine	Br	65	Terbium	Tb
5	Boron	B	36	Krypton	Kr	66	Dysprosium	Dy
6	Carbon	C	37	Rubidium	Rb	67	Holmium	Ho
7	Nitrogen	N	38	Strontium	Sr	68	Erbium	Er
8	Oxygen	O	39	Yttrium	Y	69	Thulium	Tm
9	Fluorine	F	40	Zirconium	Zr	70	Ytterbium	Yb
10	Neon	Ne	41	Niobium	Nb	71	Lutetium	Lu
11	Sodium	Na		(Columbium)		72	Hafnium	Hf
12	Magnesium	Mg	42	Molybdenum	Mo	73	Tantalum	Ta
13	Aluminum	Al	43	Technetium	Tc	74	Tungsten	W
14	Silicon	Si	44	Ruthenium	Ru	75	Rhenium	Re
15	Phosphorus	P	45	Rhodium	Rh	76	Osmium	Os
16	Sulfur	S	46	Palladium	Pd	77	Iridium	Ir
17	Chlorine	Cl	47	Silver	Ag	78	Platinum	Pt
18	Argon	A	48	Cadmium	Cd	79	Gold	Au
19	Potassium	K	49	Indium	In	80	Mercury	Hg
20	Calcium	Ca	50	Tin	Sn	81	Thallium	Tl
21	Scandium	Sc	51	Antimony	Sb	82	Lead	Pb
22	Titanium	Ti	52	Tellurium	Te	83	Bismuth	Bi
23	Vanadium	V	53	Iodine	I	84	Polonium	Po
24	Chromium	Cr	54	Xenon	Xe	85	Astatine	At
25	Manganese	Mn	55	Cesium	Cs	86	Radon	Rn
26	Iron	Fe	56	Barium	Ba	87	Francium	Fr
27	Cobalt	Co	57	Lanthanum	La	88	Radium	Ra
28	Nickel	Ni	58	Cerium	Ce	89	Actinium	Ac
29	Copper	Cu	59	Praseodymium	Pr	90	Thorium	Th
30	Zinc	Zn	60	Neodymium	Nd	91	Protactinium	Pa
31	Gallium	Ga	61	Promethium	Pm	92	Uranium	U

THE ARTIFICIAL ELEMENTS

Atomic Number	Name	Symbol	Atomic Number	Name	Symbol	Atomic Number	Name	Symbol
93	Neptunium	Np	97	Berkelium	Bk	101	Mendelevium	Mv
94	Plutonium	Pu	98	Californium	Cf	102	Nobelium	No
95	Americium	Am	99	Einsteinium	E	103	Lawrencium	Lw
96	Curium	Cm	100	Fermium	Fm			

the compound

Actually, there are many more materials than there are elements. The reason for this is that the elements can be combined to produce materials that have characteristics that are completely different from the elements. Water, for example, is a *compound* that is made from the element hydrogen and the element oxygen. And, ordinary table salt is made from the element sodium and the element chlorine.

Notice how, although hydrogen and oxygen are gases, they can produce water as a liquid.

COMPOUNDS MAY BE PRODUCED BY COMBINING:

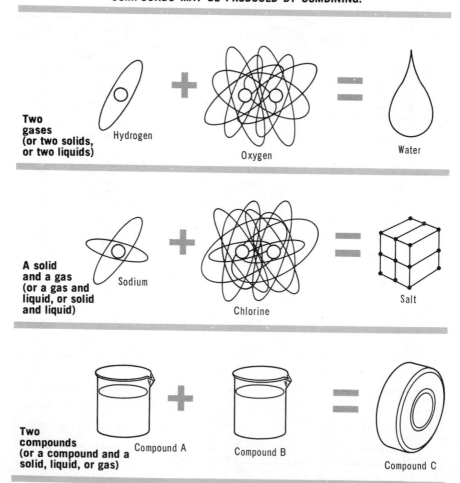

Two gases (or two solids, or two liquids)   Hydrogen   Oxygen   Water

A solid and a gas (or a gas and liquid, or solid and liquid)   Sodium   Chlorine   Salt

Two compounds (or a compound and a solid, liquid, or gas)   Compound A   Compound B   Compound C

the molecule

The molecule is the smallest particle that a *compound* can be reduced to before it breaks down into its elements. For example, if we took a grain of table salt and kept breaking it in half till it got as small as it possibly could and yet still be salt, we would have a *molecule* of salt. If we then broke it in half again, the salt would change into its elements.

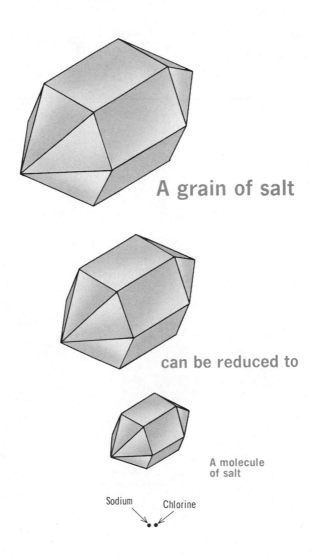

A grain of salt

can be reduced to

A molecule
of salt

Sodium   Chlorine

the atom

The atom is the smallest particle that an *element* can be reduced to and still keep the properties of that element. If a drop of water was reduced to its smallest size, a molecule of water would be produced. But if that molecule of water was reduced still further, *atoms* of hydrogen and oxygen would appear.

A molecule of water

becomes

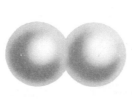

Two atoms
of hydrogen

One atom
of oxygen

This is why the chemical
formula for water is
H_2O

THE CARBON ATOM

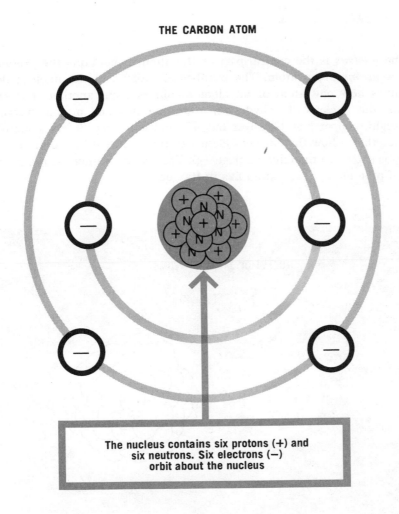

The nucleus contains six protons (+) and six neutrons. Six electrons (−) orbit about the nucleus

the structure of the atom

If the atom of an element is broken down any further, that element would no longer exist in the particles that remain. The reason for this is that these smaller particles are present in all the atoms of the different elements. The atom of one element differs from the atom of another element only because it contains different numbers of these subatomic particles.

Basically, an atom contains three types of subatomic particles that are of interest in electricity: *electrons*, *protons*, and *neutrons*. The protons and neutrons are located in the center, or *nucleus*, of the atom, and the electrons travel about the nucleus in *orbits*.

the nucleus

The *nucleus* is the central part of the atom. It contains the *protons* and *neutrons* of an atom. The number of protons in the nucleus determines how one atom of an element differs from another. For example, the nucleus of a hydrogen atom contains one proton, oxygen has eight, copper has 29, silver has 47, and gold has 79. As a matter of fact, this is how the different elements are identified by *atomic numbers,* as shown in the table on page 1-5. The atomic number is the number of protons that each atom has in its nucleus.

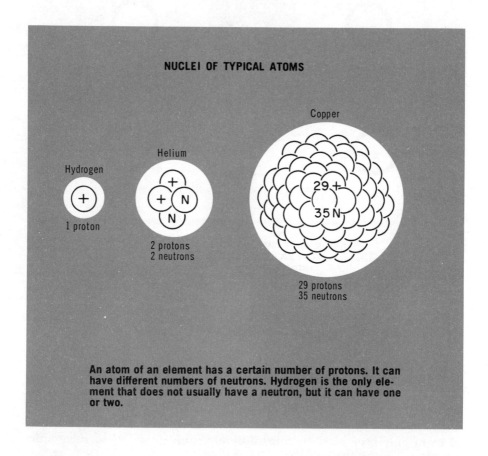

NUCLEI OF TYPICAL ATOMS

Hydrogen
1 proton

Helium
2 protons
2 neutrons

Copper
29 protons
35 neutrons

An atom of an element has a certain number of protons. It can have different numbers of neutrons. Hydrogen is the only element that does not usually have a neutron, but it can have one or two.

Although a neutron is actually a particle by itself, it is generally thought of as an electron and proton combined, and is *electrically neutral.* Since neutrons are electrically neutral, they are not too important to the electrical nature of atoms.

the proton

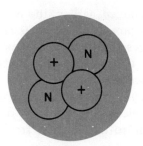

Since the nucleus of an atom contains neutrons, which are neutral, and protons, which are positive, the nucleus of any atom is always positive

The positive lines of force of a proton go straight out in all directions

The proton is very small. It is estimated to be 0.07 trillionth of an inch in diameter. The proton is one-third the diameter of an electron, but it has almost 1840 times the *mass* of an electron; the proton is almost 1840 times *heavier* than the electron. It is extremely difficult to dislodge a proton from the nucleus of an atom. Therefore, in electrical theory, protons are considered permanent parts of the nucleus. Protons do not take an active part in the flow or transfer of electrical energy.

The *proton* has a *positive* electrical *charge*. The lines of force of this charge go straight *out* in all directions from the proton.

the electron

As explained earlier, the electron is three times larger in diameter than the proton, or about 0.22 trillionth of an inch; but it is about 1840 times lighter than the proton. The electrons are *easy to move*. They are the particles that actively participate in the flow or transfer of electrical energy.

Electrons revolve in orbits around the nucleus of an atom, and have *negative* electrical *charges*. The lines of force of these charges come straight *in* to the electron from all sides.

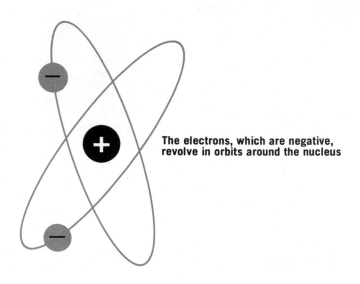

The electrons, which are negative, revolve in orbits around the nucleus

The negative lines of force of an electron come straight in from all directions

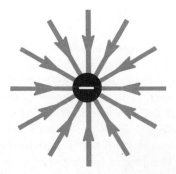

summary

☐ Electricity is produced by tiny particles called electrons and protons. ☐ Matter is anything that has weight and occupies space. It can be in the form of a solid, liquid, or gas. ☐ The basic materials that make up matter are the elements. ☐ There are ninety-two natural elements. All others are man-made. ☐ The elements can be combined to form compounds, with characteristics that are completely different from the elements from which they are produced.

☐ The molecule is the smallest particle that a compound can be reduced to before it breaks down into its elements. ☐ The atom is the smallest particle that an element can be reduced to and still keep the properties of that element. ☐ The atom of one element differs from the atom of another element only because it contains different numbers of subatomic particles. ☐ The three basic types of subatomic particles that are of interest in electricity are electrons, protons, and neutrons.

☐ The nucleus is the central part of the atom. ☐ The number of protons in the nucleus determines how the elements differ from each other. The different elements are identified by the atomic number. The atomic number is the number of protons in the nucleus. ☐ The proton has a positive charge, is smaller but 1840 times heavier than the electron, and is in the nucleus of the atom. It is difficult to dislodge from the nucleus. ☐ The electron has a negative charge, and is larger but 1840 times lighter than the proton. It revolves around the nucleus in orbits, and is easy to move. ☐ The neutron is electrically neutral, and is in the nucleus.

review questions

1. What particles produce electricity?
2. There are how many natural occurring elements?
3. Define *atomic number* of an element.
4. The proton has a _____ charge, and the electron has a _____ charge.
5. What particles are found in the nucleus of an atom? In the orbits?
6. What is the smallest particle that retains the characteristics of the compound? Of the element?
7. Is salt an element or a compound? Oxygen? Water?
8. Which is heavier, and by how much: a proton or an electron?
9. Which has a greater diameter: a proton or an electron? By how much?
10. What is the electrical charge of a neutron?

the law of electrical charges

The *negative* charge of an electron is *equal* but *opposite* to the *positive* charge of a proton.

The charges on an electron and a proton are called *electrostatic charges*. The lines of force associated with each particle produce *electrostatic fields*. Because of the way these fields act together, charged particles can attract or repel one another. The law of electrical charges is that particles with *like charges repel* each other, and those with *unlike charges attract* each other.

     A proton (+) repels another proton (+).

     An electron (−) repels another electron (−).

     A proton (+) attracts an electron (−).

Because protons are relatively heavy, the repulsive force they exert on one another in the nucleus of an atom has little effect.

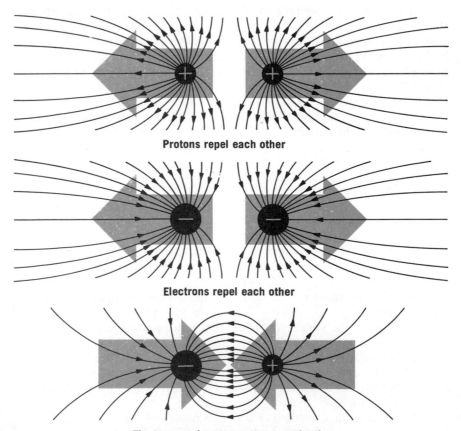

Protons repel each other

Electrons repel each other

Electrons and protons attract each other

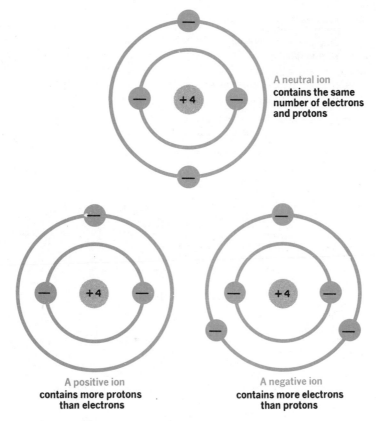

A neutral ion
**contains the same
number of electrons
and protons**

A positive ion
**contains more protons
than electrons**

A negative ion
**contains more electrons
than protons**

atomic charges

Normally, an atom contains the same number of electrons and protons, so that the *equal* and *opposite* negative and positive charges cancel each other to make the atom electrically *neutral*. But, as was explained earlier, since it is the number of protons in the nucleus that gives the atom of an element its properties, the number of electrons can be changed.

The drawings on this page show beryllium atoms, which have four protons in the nucleus. When the atom also has four electrons, the equal number of positive charges (protons) and negative charges (electrons) cancel each other, and the atom has no charge. If the beryllium atom has only three electrons, it will have more protons (+) than electrons (−), so the atom will have a *positive charge*. When the atom has five electrons, it will have more electrons (−) than protons (+). The atom will then have a *negative charge*.

Charged atoms are called *ions*. A positively charged atom is a *positive ion*, and a negatively charged atom is a *negative ion*.

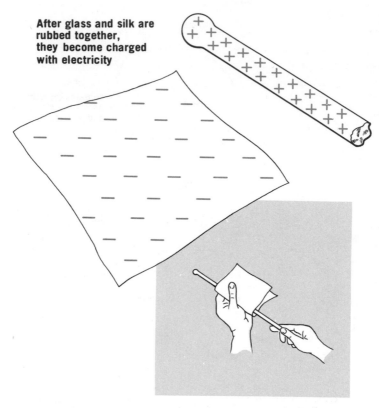

After glass and silk are
rubbed together,
they become charged
with electricity

charged materials

If a large number of atoms in a piece of neutral material loses or gains electrons, that material will become charged. Atoms can be made to do this in a number of ways, as will be explained later. The method that the ancient Greeks discovered was by *friction*. For example, if we rub a glass rod with a piece of silk, the glass rod will give up electrons to the silk. The glass rod will become positively charged, and the silk will become negatively charged.

The reason these charges result is because the glass rod has surface electrons that are easily dislodged by friction. This same thing will happen when any two materials are rubbed together, as long as one material can give up electrons easily, and the other material will accept those electrons readily. If you experiment with a few different combinations, you will find that some work well while others do not. If you comb your hair with a rubber or plastic comb, you will find that the comb will become charged, and will attract pieces of paper. Along the same lines, if you rub a rubber rod with fur, the rod will be charged negatively because it picks up electrons from the fur.

charging by contact

Suppose you had a rubber rod that was charged negatively by a piece of fur. Using this charged rubber rod, you could now *charge* other materials such as copper just by *touching* them. This method is called charging by contact, and it works because the negative charge of the rod tries to *repel electrons* from the rod's surface.

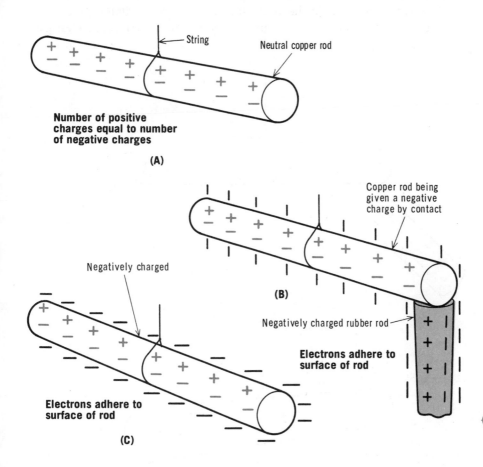

Number of positive charges equal to number of negative charges

String

Neutral copper rod

(A)

Copper rod being given a negative charge by contact

Negatively charged

(B)

Negatively charged rubber rod

Electrons adhere to surface of rod

Electrons adhere to surface of rod

(C)

The electrons on the surface of the rubber rod will go onto the surface of the *suspended* copper rod to give it a negative charge. If a positive glass rod is used instead of a negative rubber rod, electrons would be attracted from the surface of the copper rod to give it a positive charge.

charging by induction

Because electrons and protons have *attracting* and *repelling* forces, an object can be charged without being touched by the charged body. For example, if the negatively charged rubber rod is brought *close* to a piece of aluminum, the *negative* force from the rubber rod will *repel* the *electrons* in the aluminum rod to the other end. One end of the rod will then be negative, and the other positive. If we move the rubber rod away, the electrons in the aluminum rod will redistribute themselves to neutralize the rod. If you want the aluminum to remain charged,

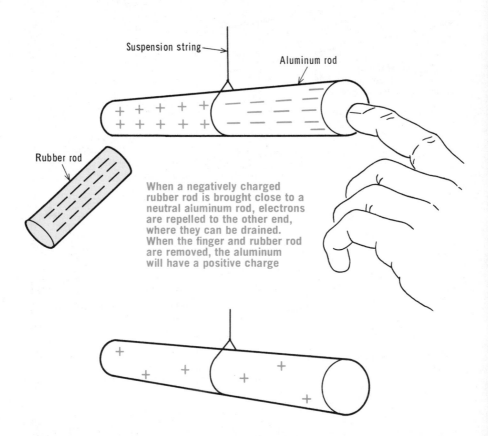

Suspension string

Aluminum rod

Rubber rod

When a negatively charged rubber rod is brought close to a neutral aluminum rod, electrons are repelled to the other end, where they can be drained. When the finger and rubber rod are removed, the aluminum will have a positive charge

bring the rubber rod close again and then touch the negative end with your finger. Electrons will leave the rod through your body. (Very small charges are involved here, and so you will not feel it.) Then, when your finger is taken away before the rubber rod is removed, the aluminum rod will remain charged. This method is called *charging by induction*.

neutralizing a charge

After glass and silk are rubbed together, they become charged with electricity. But, if the glass rod and silk are brought together again, the attraction of the positive ions in the rod pulls the electrons back out of the silk until both materials become electrically neutral.

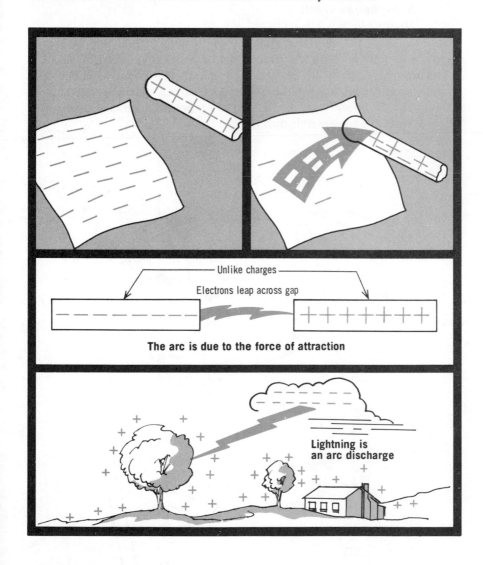

Unlike charges

Electrons leap across gap

The arc is due to the force of attraction

Lightning is an arc discharge

A wire can also be connected between the charged bodies to *discharge* them. But, if the charges on both materials are strong enough, they could discharge through an arc, like lightning.

attraction and repulsion

If you give glass rods a positive charge by rubbing them with silk, and give rubber rods a negative charge by rubbing them with fur, and then experiment with the glass, rubber, silk, and fur, without letting them touch, you would find that:

Like charges repel.

Unlike charges attract.

If you pivot the negative rubber rod to swing freely, it would be repelled by another negative rubber rod or a negative piece of silk if it were brought close to it. On the other hand, the negative rubber rod would be attracted toward the positive glass rod or the positive piece of fur. The two positive glass rods would repel each other just the way the two negative rubber rods did.

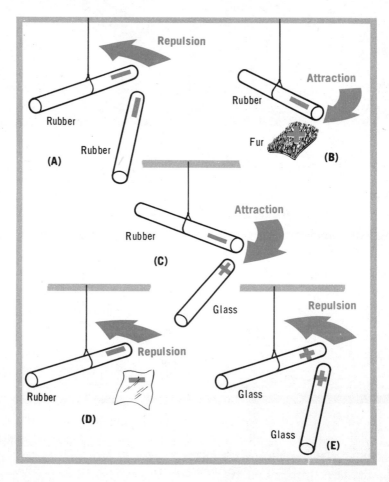

electrostatic fields

The attracting and repelling forces on charged materials occur because of the *electrostatic lines of force* that exist around the charged materials.

In a *negatively charged* object, the lines of force of the excess electrons add to produce an electrostatic field that has lines of force coming *into* the object from all directions.

In a *positively charged* object, the lack of electrons causes the lines of force on the excess protons to add to produce an electrostatic field that has lines of force going *out* of the object in all directions.

These electrostatic fields either *aid* or *oppose* each other to *attract* or *repel*.

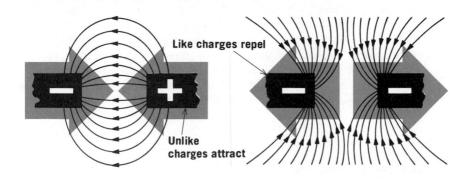

The strength of the attraction or repulsion force depends on two factors: (1) the amount of *charge* that is on each object, and (2) the distance between the objects. The greater the electric charges on the objects, the greater will be the electrostatic force. And the closer the charged objects are to each other, the greater the electrostatic force. The attraction or repulsion force gets weaker if either charge is reduced, or if the objects are moved farther apart.

During the 18th century, a scientist named Coulomb experimented with electrostatic charges and came up with a law of electrostatic attraction, which is commonly referred to as *Coulomb's law of electrostatic charges*. The law is: the force of electrostatic attraction or repulsion is directly proportional to the product of the two charges, and inversely proportional to the square of the distance between them. Of course, the more surplus electrons that a charged object has, the greater its negative charge will be. And the greater its lack of electrons, the greater its positive charge. This is explained further on page 1-56.

summary

☐ The negative electrostatic charge of an electron is equal and opposite to the positive charge of a proton. ☐ Electrostatic fields are produced by the lines of force associated with the charges. ☐ Like charges repel. A proton (+) repels another proton (+). An electron (−) repels another electron (−). ☐ Unlike charges attract. A proton (+) attracts an electron (−).

☐ An atom is neutral if it contains the same number of protons and electrons. ☐ If an atom contains less electrons than protons, it has a positive charge. ☐ If an atom contains more electrons than protons, it has a negative charge. ☐ Atoms that have a positive or negative charge are called ions. ☐ Objects can be charged by friction, contact, or induction. ☐ Neutralizing a charged object can be done by bringing it in contact with an object of opposite charge.

☐ Lines of force enter a negatively charged object, by convention. ☐ Lines of force leave a positively charged object, by convention. ☐ The electrostatic fields created by the lines of force aid or oppose each other to attract or repel. ☐ The strength of the attraction or repulsion force depends on the amount of charge that is on each object, and the distance between the objects. ☐ Coulomb's Law of Electrostatic Charges relates the forces of attraction and repulsion: Force is directly proportional to the product of the two charges, and inversely proportional to the square of the distance between them.

review questions

1. If an electron were brought in the vicinity of a proton, would the proton repel or attract the electron?
2. Why don't protons in a nucleus repel each other with sufficient force to split the nucleus?
3. Do the protons in a nucleus have any repulsive force on each other?
4. What is the polarity of the charge of an object that has less electrons than protons?
5. Name three ways in which a material can be charged.
6. If a rubber rod is rubbed with a piece of fur, what is the polarity of the rubber? What is the polarity of the fur?
7. How can a charged object be neutralized?
8. Do the lines of force enter or leave an electron?
9. Does the force of repulsion between two electrons increase or decrease with distance? If the distance is doubled, what is the magnitude of the new force as compared to the old?
10. State Coulomb's Law.

electron orbits

As you have seen, electricity is produced when electrons leave their atoms. To understand the different ways of accomplishing this, it would be helpful to know more about the nature of the different *electron orbits* about the nucleus of an atom.

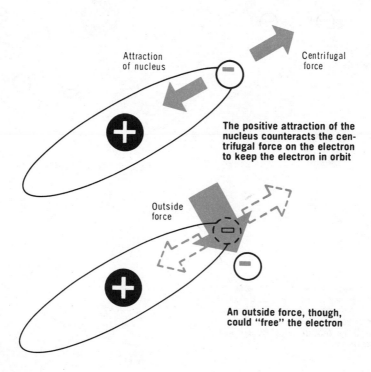

Attraction of nucleus

Centrifugal force

The positive attraction of the nucleus counteracts the centrifugal force on the electron to keep the electron in orbit

Outside force

An outside force, though, could "free" the electron

The electrons *revolve* at high speed in their orbits about the atom's nucleus. Because of the electron's great speed, centrifugal force tends to pull the electron out of orbit. But the *positive attraction* of the *nucleus* keeps the electron from breaking away. However, if a sufficient outside force were applied to aid the centrifugal force, the electron could be "freed."

When a force is applied to an atom, it gives energy to the orbiting electron, and it is the amount of energy absorbed by the electron that determines whether or not it will be freed. There are a number of ways that energy can be applied, which you will learn about later: friction, chemicals, heat, pressure, magnetism, and light.

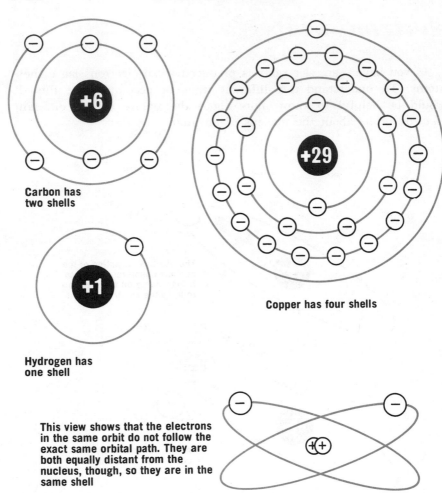

Carbon has
two shells

Copper has four shells

Hydrogen has
one shell

This view shows that the electrons
in the same orbit do not follow the
exact same orbital path. They are
both equally distant from the
nucleus, though, so they are in the
same shell

orbital shells

Electrons that orbit close to the nucleus are difficult to free because
they are close to the positive force that holds them. The farther the
electrons are from the nucleus, the weaker the positive force becomes.
As you might have noticed in some of the earlier diagrams, the more
electrons that an atom has, the more orbits there are. The orbital paths
are commonly called *shells*.

The atoms of all the known elements can have up to seven shells.
The table on page 1-25 lists the 103 elements, showing the number
of electrons in each shell for each atom.

the elements
and their atomic shells

ELECTRON SHELLS

Atomic No.	Element	Electrons per shell					Atomic No.	Element	Electrons per shell						
		1	2	3	4	5			1	2	3	4	5	6	7
1	Hydrogen, H	1					53	Iodine, I	2	8	18	18	7		
2	Helium, He	2					54	Xenon, Xe	2	8	18	18	8		
3	Lithium, Li	2	1				55	Cesium, Cs	2	8	18	18	8	1	
4	Beryllium, Be	2	2				56	Barium, Ba	2	8	18	18	8	2	
5	Boron, B	2	3				57	Lanthanum, La	2	8	18	18	9	2	
6	Carbon, C	2	4				58	Cerium, Ce	2	8	18	19	9	2	
7	Nitrogen, N	2	5				59	Praseodymium, Pr	2	8	19	20	9	2	
8	Oxygen, O	2	6				60	Neodymium, Nd	2	8	19	21	9	2	
9	Fluorine, F	2	7				61	Promethium, Pm	2	8	18	22	9	2	
10	Neon, Ne	2	8				62	Samarium, Sm	2	8	18	23	9	2	
11	Sodium, Na	2	8	1			63	Europium, Eu	2	8	18	24	9	2	
12	Magnesium, Mg	2	8	2			64	Gadolinium, Gd	2	8	18	25	9	2	
13	Aluminum, Al	2	8	3			65	Terbium, Tb	2	8	18	26	9	2	
14	Silicon, Si	2	8	4			66	Dysprosium, Dy	2	8	18	27	9	2	
15	Phosphorus, P	2	8	5			67	Holmium, Ho	2	8	18	28	9	2	
16	Sulfur, S	2	8	6			68	Erbium, Er	2	8	18	29	9	2	
17	Chlorine, Cl	2	8	7			69	Thulium, Tm	2	8	18	30	9	2	
18	Argon, A	2	8	8			70	Ytterbium, Yb	2	8	18	31	9	2	
19	Potassium, K	2	8	8	1		71	Lutetium, Lu	2	8	18	32	9	2	
20	Calcium, Ca	2	8	8	2		72	Hafnium, Hf	2	8	18	32	10	2	
21	Scandium, Sc	2	8	9	2		73	Tantalum, Ta	2	8	18	32	11	2	
22	Titanium, Ti	2	8	10	2		74	Tungsten, W	2	8	18	32	12	2	
23	Vanadium, V	2	8	11	2		75	Rhenium, Re	2	8	18	32	13	2	
24	Chromium, Cr	2	8	13	1		76	Osmium, Os	2	8	18	32	14	2	
25	Manganese, Mn	2	8	13	2		77	Iridium, Ir	2	8	18	32	15	2	
26	Iron, Fe	2	8	14	2		78	Platinum, Pt	2	8	18	32	16	2	
27	Cobalt, Co	2	8	15	2		79	Gold, Au	2	8	18	32	18	1	
28	Nickel, Ni	2	8	16	2		80	Mercury, Hg	2	8	18	32	18	2	
29	Copper, Cu	2	8	18	1		81	Thallium, Tl	2	8	18	32	18	3	
30	Zinc, Zn	2	8	18	2		82	Lead, Pb	2	8	18	32	18	4	
31	Gallium, Ga	2	8	18	3		83	Bismuth, Bi	2	8	18	32	18	5	
32	Germanium, Ge	2	8	18	4		84	Polonium, Po	2	8	18	32	18	6	
33	Arsenic, As	2	8	18	5		85	Astatine, At	2	8	18	32	18	7	
34	Selenium, Se	2	8	18	6		86	Radon, Rn	2	8	18	32	18	8	
35	Bromine, Br	2	8	18	7		87	Francium, Fr	2	8	18	32	18	8	1
36	Krypton, Kr	2	8	18	8		88	Radium, Ra	2	8	18	32	18	8	2
37	Rubidium, Rb	2	8	18	8	1	89	Actinium, Ac	2	8	18	32	18	9	2
38	Strontium, Sr	2	8	18	8	2	90	Thorium, Th	2	8	18	32	19	9	2
39	Yttrium, Y	2	8	18	9	2	91	Protactinium, Pa	2	8	18	32	20	9	2
40	Zirconium, Zr	2	8	18	10	2	92	Uranium, U	2	8	18	32	21	9	2
41	Niobium, Nb	2	8	18	12	1	93	Neptunium, Np	2	8	18	32	22	9	2
42	Molybdenum, Mo	2	8	18	13	1	94	Plutonium, Pu	2	8	18	32	23	9	2
43	Technetium, Tc	2	8	18	14	1	95	Americium, Am	2	8	18	32	24	9	2
44	Ruthenium, Ru	2	8	18	15	1	96	Curium, Cm	2	8	18	32	25	9	2
45	Rhodium, Rh	2	8	18	16	1	97	Berkelium, Bk	2	8	18	32	26	9	2
46	Palladium, Pd	2	8	18	18	0	98	Californium, Cf	2	8	18	32	27	9	2
47	Silver, Ag	2	8	18	18	1	99	Einsteinium, E	2	8	18	32	28	9	2
48	Cadmium, Cd	2	8	18	18	2	100	Fermium, Fm	2	8	18	32	29	9	2
49	Indium, In	2	8	18	18	3	101	Mendelevium, Mv	2	8	18	32	30	9	2
50	Tin, Sn	2	8	18	18	4	102	Nobelium, No	2	8	18	32	31	9	2
51	Antimony, Sb	2	8	18	18	5	103	Lawrencium, Lw	2	8	18	32	32	9	2
52	Tellurium, Te	2	8	18	18	6									

shell capacity

If you study the table on page 1-25 briefly, you will notice that each shell can only hold a certain number of electrons. The shell closest to the nucleus (the first shell) cannot hold more than 2 electrons; the second shell cannot hold more than 8 electrons; the third, no more than 18; the fourth, no more than 32; and so on.

If you look again at the table on page 1-25, you will see that up to atomic number 10, the second shell built up to 8 electrons. Since this is the limit for the second shell, a third shell had to be started. From atomic numbers 11 through 18, the third shell built up to 8, and then a fourth shell started. Then, from numbers 19 through 29, the third shell built up to its maximum of 18.

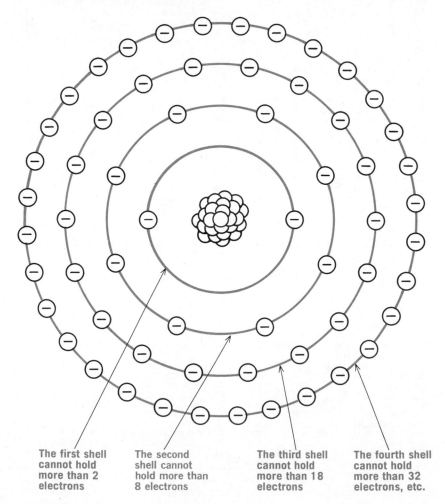

The first shell cannot hold more than 2 electrons

The second shell cannot hold more than 8 electrons

The third shell cannot hold more than 18 electrons

The fourth shell cannot hold more than 32 electrons, etc.

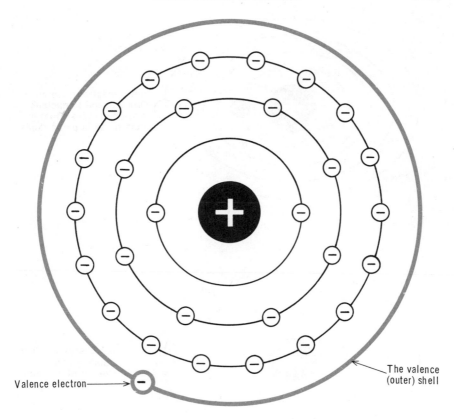

**The outer shell is called the valence shell, and the
electrons in that shell are called valence electrons**

the outer (valence) shell

As you can notice in the table on page 1-25, although the third shell can hold up to 18 electrons, it did not take on any more than 8 electrons until the fourth shell started. This is also true of the fourth shell. It will not take on any more electrons than 8 until a fifth shell starts, even though the fourth shell can hold up to 32 electrons. This shows that there is another rule. *The outer shell of an atom will have no more than 8 electrons.* The outer shell of an atom is called the *valence shell,* and its electrons are called *valence electrons.* The number of electrons in the valence shell of an atom is important in electricity, as you will see later.

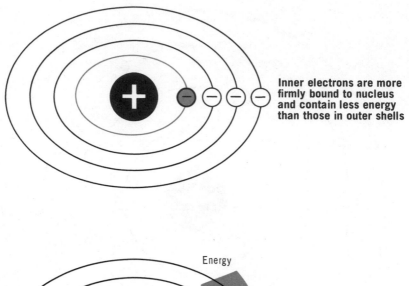

Inner electrons are more firmly bound to nucleus and contain less energy than those in outer shells

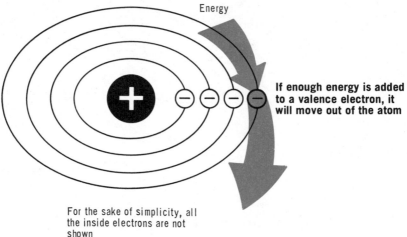

Energy

If enough energy is added to a valence electron, it will move out of the atom

For the sake of simplicity, all the inside electrons are not shown

electron energy

Although every electron has the same negative charge, not all electrons have the same *energy level*. The electrons that orbit close to the nucleus contain less energy than those that orbit further away. The further an electron orbits from the nucleus, the greater its energy.

If enough energy is added to an electron, that electron will move *out of its orbit* to the next higher orbit. And, if enough energy is added to a *valence* electron, that electron will move *out of its atom,* since there is no next higher orbit.

producing electricity

Electricity is produced when *electrons are freed* from their atoms. Since the valence electrons are farthest from the attractive force of the nucleus and also have the highest energy level, they are the electrons that are most easily set free. When enough force or energy is applied to an atom, the valence electrons will become free. However, the energy supplied to a valence shell is distributed amongst the electrons in that shell. Therefore, for a given amount of energy, the more valence electrons there are, the *less* energy each electron will get.

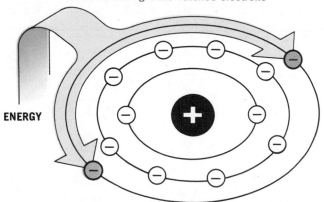

Energy is applied to the valence shell, and is distributed amongst the valence electrons

Two electrons share the energy equally

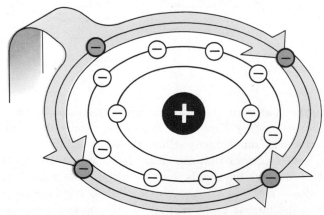

Four electrons share the energy equally, but each has less energy gained than any electron above

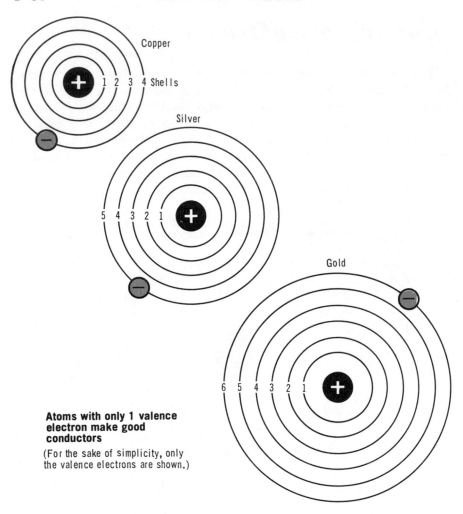

Copper

1 2 3 4 Shells

Silver

5 4 3 2 1

Gold

6 5 4 3 2 1

**Atoms with only 1 valence
electron make good
conductors**

(For the sake of simplicity, only
the valence electrons are shown.)

conductors

The valence shell can contain up to 8 valence electrons. Since valence electrons share any energy applied to them, the atoms that have less valence electrons will more easily allow those electrons to be freed. Materials that have electrons that are easily freed are called *conductors*. The atoms of conductors have only 1 or 2 valence electrons. The ones with only *1 valence electron* are the best electrical conductors.

If you look at the atomic table on page 1-25, you can pick the good conductors. They all have one electron in their outer shell. Most *metals* are good conductors. The ones you are probably most familiar with are: copper (No. 29), silver (No. 47), and gold (No. 79).

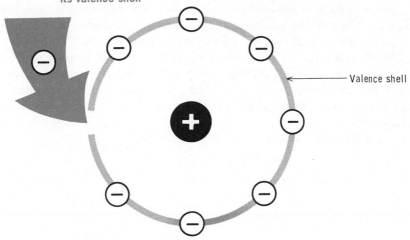

An atom that is more than half filled, but has less than 8 electrons, tries to become stable by filling its valence shell

Valence shell

These atoms make good insulators because it is very difficult to free an electron from their valence shell

insulators

Insulators are materials from which electrons are very difficult to free. The atoms of insulators have their valence shells filled with 8 electrons or are more than half filled. Any energy applied to such an atom will be distributed amongst a relatively large number of electrons. But, in addition to this, these atoms resist giving up their electrons because of a phenomenon known as *chemical stability*.

An atom is completely *stable* when its *outer shell* is completely *filled* or when it has 8 valence electrons. A stable atom resists any sort of activity. In fact, it will not combine with any other atoms to form compounds. There are six naturally stable elements: helium, neon, argon, krypton, xenon, and radon. These are known as the *inert gases*.

All atoms that have less than 8 valence electrons tend to attain the stable state. Those that are less than half filled (the conductors), tend to release their electrons to empty the unstable shell. But those that are more than half filled (the insulators), strive to collect electrons to fill up the valence shell. So, not only is it difficult to free their electrons, but these atoms of the insulators also oppose the production of electricity with their tendency to catch any electrons that may be freed. Those atoms that have 7 valence electrons most actively try to be filled, and are excellent electrical insulators.

semiconductors

Semiconductors are those materials that are neither good conductors nor good insulators. In other words, they can conduct electricity better than insulators can, but not as well as conductors can.

Semiconductors do not conduct electricity as well as good conductors because they have more than 1 or 2 valence electrons, and so do not give up the electrons so easily. By the same token, semiconductors have less than 7 or 8 valence electrons, which is what insulators have. Thus, they do not resist giving up electrons as much as the insulators do, and therefore will allow some conduction.

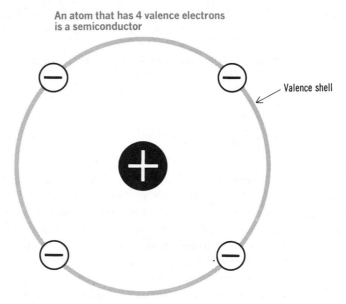

An atom that has 4 valence electrons
is a semiconductor

Valence shell

These atoms give up electrons less easily
than good conductor atoms, but more easily
than good insulator atoms

Generally, good conductors have their valence shells less than half filled, and good insulators have their valence shells more than half filled. So, a semiconductor is a material whose atoms are *half filled*. Those atoms have 4 valence electrons.

A review of the table on page 1-25 shows that carbon, silicon, germanium, tin, and lead are all semiconductors, since they have 4 electrons in their outer (valence) shell.

atomic bonds

Until now, you have studied the characteristics of individual atoms in relation to conductors, semiconductors, and insulators. Actually, when you work with materials, the characteristics of the *molecules* are what become important. As you recall, atoms join together to form molecules. Because of the nature of the way atoms bond together, the characteristics of the molecules they form can be quite different from those of the atoms that form the molecule.

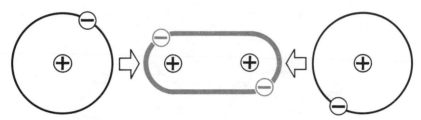

Two hydrogen atoms join together when each of them shares its electrons with the other to form a stable hydrogen molecule

You remember that atoms tend to strive for *chemical stability*, the condition in which the outer or valence shell is completely filled. Atoms, then, tend to bond together in such a way that this stability is brought about.

Hydrogen is a good example of the simple way this happens. As the table on page 1-25 shows, the hydrogen atom has only one shell and one electron. The table also shows that the maximum number of electrons this first shell can hold is 2. Thus, the hydrogen atoms will combine to become stable with each atom having 2 valence electrons. This is done by the two atoms joining together and *sharing* their electrons. So, even though there is a total of only 2 electrons for both atoms, when the *electron pair is shared*, each atom sees 2 electrons, and a stable molecule is produced. This is known as *electron pair* or *covalent bonding*.

Other atoms join in similar ways, with each determined by the number needed for stability. Take chlorine, for example, which has 7 valence electrons. As the table on page 1-25 shows, 8 electrons are needed in its valence shell for stability. If two atoms each permit one electron to be shared as a pair, each atom in the completed molecule will have 8 valence electrons. For oxygen, which has 6 valence electrons, three atoms will join, with one forming an *electron pair bond* with each

atomic bonds (cont.)

of the other two atoms. You can see, then, that the number of atoms that must join together to form a *stable molecule* depends on the particular atoms involved. With atoms that have fewer valence electrons, the molecular structure becomes more and more complex.

On the other hand, those atoms of elements whose valence shells are already filled do not have to join with other atoms. In these elements, the atoms *are* the molecules.

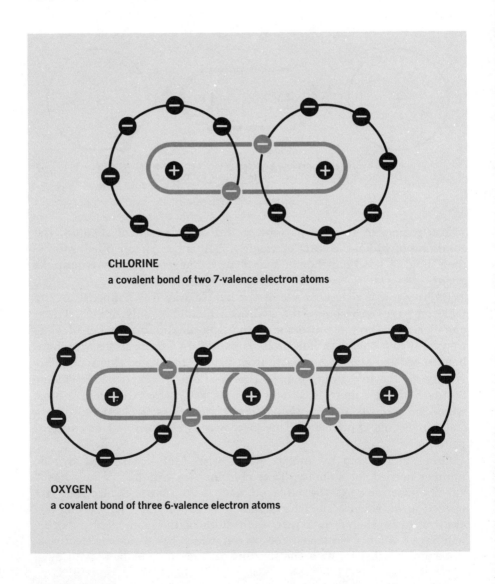

CHLORINE
a covalent bond of two 7-valence electron atoms

OXYGEN
a covalent bond of three 6-valence electron atoms

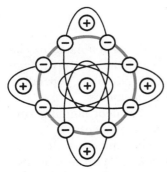

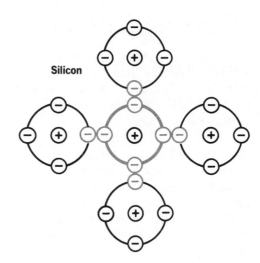

Silicon

Each of the four atoms shares an electron with the center atom, so that only the center atom is stable with 8. This is how the five atoms look with the electrons orbiting

covalent bonds

Semiconductor materials are a good example of how atoms must join together in a complex way to produce a stable molecular arrangement. Silicon, one of the better-known semiconductors, has atoms with 4 valence electrons: 8 are needed for stability. This means that each valence electron of any atom must pair up and be shared with another valence electron from another atom. Therefore, a total of five atoms is needed to make only one of them stable. This arrangement could be considered a molecule, but actually it is not, because the other four atoms are not stable. Each of those atoms will have to join with others in the same way to become stable, but in each case there will still be some unstable atoms that must join with others. A continuous atomic structure must be formed, which is known as a *crystal lattice*. There are no real discrete molecules, but there are groups of atoms that can be considered as molecular groupings.

Covalent bonds cause the atomic structure of semiconductors to repeat the same geometric pattern throughout the material— the crystal lattice

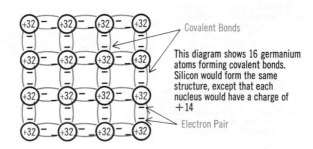

Covalent Bonds

This diagram shows 16 germanium atoms forming covalent bonds. Silicon would form the same structure, except that each nucleus would have a charge of +14

Electron Pair

electrovalent bonds

The previous pages discussed bonding primarily as it relates to *elements*. When the atoms of different elements combine to form *compounds*, the bonds can be similar and also quite different, depending on the number of valence electrons in the different atoms.

Water is an example of a compound that uses covalent bonding similar to the ordinary bonds of the atoms of the same elements. However, since two different atoms are involved with a compound, with different amounts of valence electrons, it appears different. With water, two hydrogen atoms combine with one oxygen atom to form a water molecule. Oxygen has 6 valence electrons, and hydrogen has 1. Each hydrogen electron pairs up with an oxygen electron, so that the hydrogen atoms each stabilize with an *electron pair*, and the oxygen atom sees 8.

Another form of bonding that is common with compounds is *electrovalent bonding*.

The combination of the elements sodium and chlorine is an example of this. Sodium has 1 valence electron, which it tends to give up to become stable; and chlorine has 7 valence electrons, and needs 8 to become stable. When the atoms of sodium and chlorine are brought together, the sodium atom gives up its electron to the chlorine atom. The sodium atom then becomes a *positive ion* and the chlorine atom becomes a *negative ion*. They attract each other and bond together to form sodium chloride. Because ionic attraction forms the bond, this is also known as *ionic bonding*.

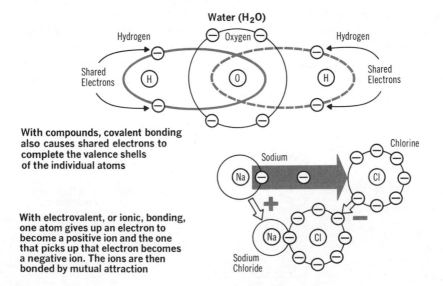

With compounds, covalent bonding also causes shared electrons to complete the valence shells of the individual atoms

With electrovalent, or ionic, bonding, one atom gives up an electron to become a positive ion and the one that picks up that electron becomes a negative ion. The ions are then bonded by mutual attraction

metallic bonds

In the strictest sense, the *metallic bond* of atoms is a form of covalent bonding, wherein the valence electrons are *shared* by the various atoms that join together. However, the covalent bonds you just studied dealt with atoms whose valence shells were more than half filled, and shared *electron pairs* to complete a stable bond.

With metals, which are good conductors, the atoms have only 1 valence electron. And these atoms seem to *lose* their valence electrons to attain stability. With elements such as copper, silver, gold, etc., the atoms join together, with each atom continually giving up its electron to a neighboring atom and taking on another electron from another neighboring atom. This process of giving up and taking on electrons is continuous, and the individual valence electrons are *free* to wander aimlessly and randomly from atom to atom in and out of orbits. This does not mean that the electrons wander free of the atoms. It means only that groups of atoms continuously *share* all of their valence electrons.

With metallic bonding, the valence electrons move from atom to atom, so that all the atoms share their valence electrons

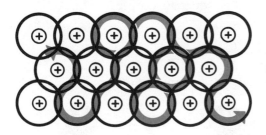

The electrons are free to wander randomly from atom to atom

Another way to look at this is that any group of atoms can form a molecular structure with all their valence electrons orbiting all of the atoms in the group. This is described as a *metallic lattice*, as opposed to the crystal lattices formed by *electron pair* bonds. Some theories describe the metallic sharing of valence electrons as producing an *electron cloud* which surrounds their bonded atoms. As you can see, there are no discrete molecules formed by metallic bonding.

the effect of atomic bonds

Earlier you learned how the number of valence electrons in the atoms of different elements determined whether those atoms made good conductors, insulators, or semiconductors, depending on the stability of the element. But now you have learned that when the atoms bond together, the valence structure of the material can change. The table on page 1-25 shows that there are more elements available with these valence shells less than half filled, which could mean that most elements are good conductors. But this is not true because of the way the atoms of most elements are bonded. Most covalent bonds produce stable atomic arrangements or molecules. Except for those elements that use metallic bonds, most elements are not good conductors. As a matter of fact, if the elements were pure, most of them would be perfect insulators.

It is important to keep in mind, however, that under practical conditions most materials contain impurities. Also, stable bonds can be broken by heat energy, so that conduction can usually take place.

Compounds are good examples of how bonds can change electrical characteristics. Copper, which is a good conductor, becomes a good insulation when combined with oxygen to produce a stable copper oxide molecule, with 8 valence electrons.

Those good conductors that use metallic bonding also end up with the different materials having different conduction characteristics. Although they all have 1 valence electron in their atoms, some metals have a denser grouping of atoms, and so have more valence electrons available in a given amount of material. This is why copper is a better conductor than gold, and silver is better than copper.

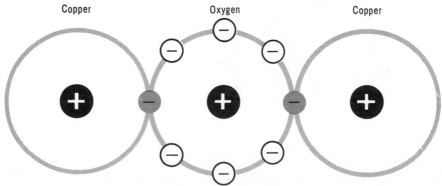

Copper is a good conductor because it has only one valence electron. But when two copper atoms combine with one oxygen atom, they produce a molecule of copper oxide (Cu_2O); eight valence electrons now exist to make the molecule stable. Copper oxide, then, is a good insulator in its pure form

summary

☐ Electrons revolve at high speed in their orbits about the atom's nucleus. This high speed gives rise to a centrifugal force that tends to pull the electron out of orbit. The positive attraction of the nucleus prevents this. A large outside force can free the electron from the atom. ☐ The positive attractive force is greater on the electrons in orbit that are closer to the nucleus; hence, they are more difficult to free.

☐ Electrons orbit in one of seven shells. ☐ The innermost shell can hold a maximum of 2 electrons; the second, 8; the third, 18; the fourth, 32; etc. ☐ The outermost shell of an atom is the valence shell. ☐ All electrons in this outer shell are called valence electrons. ☐ The valence shell never contains more than 8 electrons. ☐ An atom with a completely filled valence shell is very stable, and chemically inactive. ☐ Electrons in orbit farther away from the nucleus contain more energy. If sufficient energy is added to an electron, it will move to the next outer orbit; if sufficient energy is added to a valence electron, it will be freed. ☐ The flow of free electrons constitutes electric current.

☐ Conductors are materials that have one or two valence electrons, which are easily freed. ☐ Insulators are materials that have five or more valence electrons, which are difficult to free. ☐ Semiconductors are materials that have more free electrons than insulators, but less than conductors. Impurities can make them better conductors. ☐ Bonding is the force that keeps elements together to form compounds. The electrons in the compounds form stable octets.

review questions

1. What prevents an electron from being freed due to its centrifugal force?
2. What is a shell, and how many shells are there?
3. What is a "valence" electron?
4. What is the maximum number of valence electrons in an atom?
5. What is a "free" electron?
6. Why are compounds good insulators? Name two good insulators.
7. Name two semiconductors, and two conductors.
8. What are the characteristics of semiconductors?
9. Why are impurities added to compounds?
10. Is an element that contains six valence electrons a good conductor? Two valence electrons?

how electricity is produced

Until now, the discussion dealt with the general idea of applying a force or energy to the electrons to move them out of their orbits, but no mention was made as to how this can be done. It can be done in a variety of ways, all of which fall into six accepted categories.

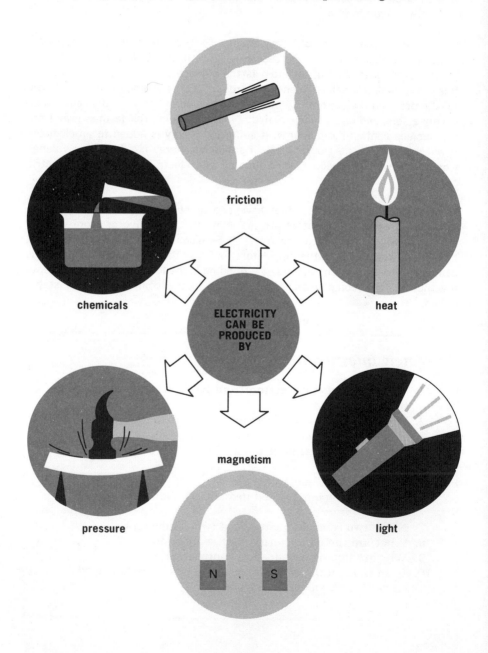

friction

heat

chemicals

ELECTRICITY CAN BE PRODUCED BY

light

pressure

magnetism

electricity from friction

This is the method that was first discovered by the ancient Greeks, which was described earlier in the book. An electric charge is produced when two pieces of material are rubbed together, such as silk and a glass rod, or when you comb your hair. Did you ever walk across a carpet and get a shock when you touched a metal doorknob? Your shoe soles built up a charge by rubbing on the carpet, and this charge was transferred to you and was discharged to the knob. These charges are called *static electricity;* and results when one material *transfers* its *electrons* to another.

TRIBOELECTRICITY

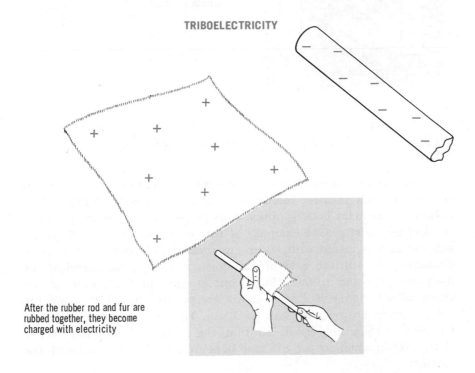

After the rubber rod and fur are rubbed together, they become charged with electricity

It is something that is still not completely understood. But, one theory is: On the *surface* of a material, there are many atoms that cannot combine with other atoms as they do inside the material; the surface atoms, then, have some free electrons. This is the reason why insulators such as glass and rubber can produce charges of static electricity. The heat energy produced by the rubbing friction is supplied to the surface atoms to release the electrons. This is known as the *triboelectric* effect.

ELECTROCHEMISTRY

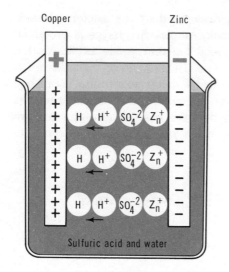

Copper   Zinc

The "Wet" Cell

The solution, known as the electrolyte, pulls positive ions from the zinc bar and free electrons from the copper bar

The "dry" flashlight battery uses an electrolytic paste instead of a fluid solution

electricity from chemicals

Chemicals can be combined with certain metals to cause a chemical action that will transfer electrons to produce electric charges. This is how the ordinary *battery* works. This process works on the principle of *electrochemistry*. One example of this is the basic *wet cell*. When sulfuric acid is mixed with water (to form the *electrolyte*) in a glass container, the sulfuric acid breaks down into separate chemicals of hydrogen (H) and sulfate (SO_4), but because of the nature of the chemical action, the hydrogen atoms are positive ions (H^+) and the sulfate atoms are negative ions (SO_4^{-2}). The number of positive and negative charges are equal, so that the entire solution has no net charge. Then when copper and zinc bars are added, they *react* with the solution.

The zinc combines with the sulfate atoms; and since those atoms are negative, positive zinc ions (Zn^+) are given off by the zinc bar. The electrons from the zinc ions are left behind in the zinc, so that the zinc bar has a surplus of electrons, or a *negative charge*. The zinc ions combine with and neutralize the sulfate ions so that the solution now has more positive charges. The positive hydrogen ions attract free electrons from the copper bar to again neutralize the solution. But, now the *copper* bar has a lack of electrons, so it gets a *positive charge*.

Batteries and cells are covered in greater detail in Volume 6.

electricity from pressure

When pressure is applied to some materials, the *force* of the *pressure* is passed through the material to its atoms, where it drives the electrons out of orbit in the direction of the force. The electrons leave one side of the material and accumulate on the other side. Thus, positive and negative charges are built up on opposite sides. When the pressure is released, the electrons return to their orbits. The materials are cut in certain ways to control the surfaces that will be charged. Some materials will react to a *bending* pressure, while others will respond to a *twisting* pressure.

Piezoelectricity is the name given to the effect of *pressure* in causing electrical charges. *Piezo* is a name derived from the Greek word for pressure. This effect is most noticeable in crystal materials, such as Rochelle salts and certain ceramics like barium titanate. These piezo crystals are used in some microphones and phonograph pickups.

PIEZOELECTRICITY

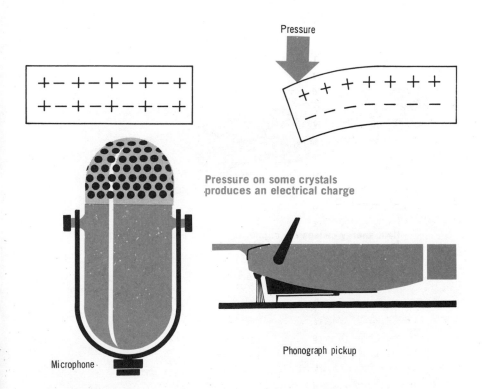

Pressure

Pressure on some crystals produces an electrical charge

Phonograph pickup

Microphone

electricity from heat

Because some materials readily give up their electrons, and other materials accept electrons, a transfer of electrons can take place when two *dissimilar metals*, for example, are joined. With particularly active metals, the heat energy of normal room temperature is enough to make those metals release electrons. Copper and zinc, for example, will act this way. Electrons will leave the copper atoms and enter the zinc atom. The zinc, then, gets a surplus of electrons and becomes negatively charged. The copper, having lost electrons, takes on a positive charge.

The charges developed at room temperature are small, though, because there is not enough heat energy to free more than a relatively few electrons. But, if heat is applied to the junction of the two metals to provide more energy, more electrons will be freed. This method is called *thermoelectricity*. The more heat that is applied, the greater the charge that will be built up. When the heat is removed, the metals will cool, and the charges will dissipate. The device described is called a *thermocouple*. When a number of thermocouples are joined together, a *thermopile* is made.

THERMOELECTRICITY

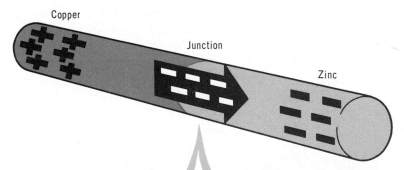

Copper

Junction

Zinc

Heat energy causes the copper to release electrons to the zinc

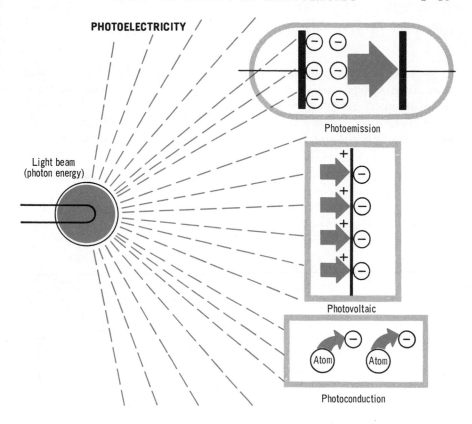

PHOTOELECTRICITY

Photoemission

Photovoltaic

Light beam
(photon energy)

Photoconduction

electricity from light

Light in itself is a form of energy, and is generally considered by many scientists to be made of small particles of energy called *photons*. When the photons in a light beam strike a material, they release their energy. With some materials, the energy from the photons can cause atoms to release electrons. Materials such as potassium, sodium, cesium, lithium, selenium, germanium, cadmium, and lead sulfide react this way to light. This *photoelectric effect* can be used in three ways:

1. *Photoemission:* The photon energy from a beam of light could cause a surface to release electrons in a vacuum tube. A plate would then collect the electrons.
2. *Photovoltaic:* The light energy on one of two plates that are joined together causes one plate to release electrons to the other. The plates build up opposite charges, like a battery.
3. *Photoconduction:* The light energy applied to some materials that are normally poor conductors causes free electrons to be produced in the materials, so that they become better conductors.

electricity from magnetism

Most of you are probably familiar with magnets, and have toyed with them at one time or another. You may have seen that, in some cases, magnets attract each other, and in other cases, they repel each other. The reason for this is that magnets have force fields that interact with one another. (This is explained more fully later in this book.)

ELECTROMAGNETISM

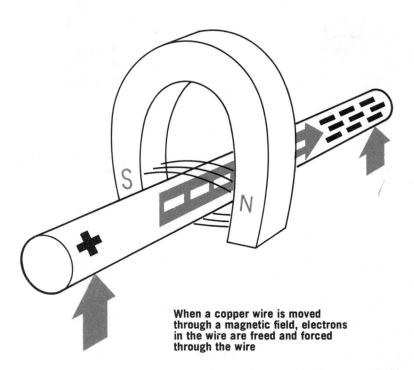

When a copper wire is moved through a magnetic field, electrons in the wire are freed and forced through the wire

The force of a magnetic field can also be used to move electrons. This is known as *magnetoelectricity,* and is the basis for how an electric *generator* produces electricity. When a good conductor, such as copper, is passed through a *magnetic* field, the *force* of the field will provide enough energy to cause the copper atoms to free their valence electrons. The electrons will all be moved in a certain direction, depending on how the wire crosses the magnetic field. Actually it is not necessary to only move the conductor through the magnetic field; the same effect will be obtained if the field is moved across the conductor. All that is really required is relative motion between any conductor and a magnetic field. (Magnetoelectricity is further examined at the end of the book.)

summary

☐ Electrons can be made to move out of their orbits by applying a force or energy to them. ☐ Six accepted ways of doing this are triboelectric effect, electrochemistry, piezoelectricity, thermoelectricity, photoelectric effect, and magnetoelectricity.

☐ Triboelectric effect causes electrons that are on the surface of a material to be released by rubbing. The energy comes from heat friction. ☐ Electrochemistry enables chemicals to be combined with certain metals to cause a chemical action that will transfer electrons to produce charges. ☐ Piezoelectricity is the effect of pressure causing electrical charges. Its effects are most noticeable in crystals. ☐ Thermoelectricity is the effect of heat being applied to two dissimilar metals, producing opposite charges on the two metals. ☐ Photoelectric effect causes the atoms of certain materials to release electrons when light energy, in the form of photons, strikes them.

☐ Photoemission: Photon energy releases electrons of one surface in a vacuum tube. Another surface in the tube collects the electrons. ☐ Photovoltaic: Light energy on one of two joined plates causes electrons to be released to the other. The plates then act like a battery. ☐ Photoconduction: Light energy applied to some materials causes them to become better conductors. ☐ Magnetoelectricity is the effect of the force of a magnetic field, which can be used to move electrons.

review questions

1. What causes electrons to move out of their orbits?
2. What is the effect of applying a pressure to a Rochelle salt crystal?
3. What is the triboelectric effect?
4. In thermoelectricity, heat is applied at the junction of two _____ metals.
5. What is the difference between a thermocouple and a thermopile?
6. What are photons?
7. Name and describe three ways in which the photoelectric effect can be used.
8. Is it always necessary for the conductor to move in order for its electrons to be freed by a magnetic force?
9. Describe the basic wet cell. It operates on the principle of _____.
10. The electric generator operates on the principle of _____.

what is electric current?

The book, till now, dealt with what electricity is, and how electric charges are produced. The subjects mostly covered what is called *static electricity,* which is an electric *charge at rest.* But, a static electric charge cannot usually perform a useful function. In order to use electrical energy to do some kind of work, the electricity must be set in "motion." This is done when an electric current is produced. This electric current is developed when many *free electrons* in a wire are moved in the same direction.

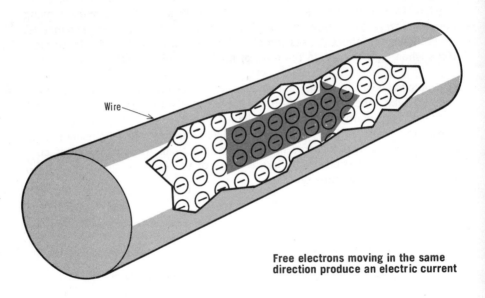

Free electrons moving in the same direction produce an electric current

As you will learn later, every electron contains energy that can cause certain effects. Ordinarily, electrons are moving in various directions, so that their effects cancel. But when we get the electrons to move in the same direction, that is, form a current flow, their effects can add, and the energy they release can be put to work. Also, the more electrons there are moving in the same direction, the greater is the current flow, and the more electron energy is made available to do work. Therefore, the larger or smaller electric currents are caused by the greater or fewer number of electrons set in motion in the same direction.

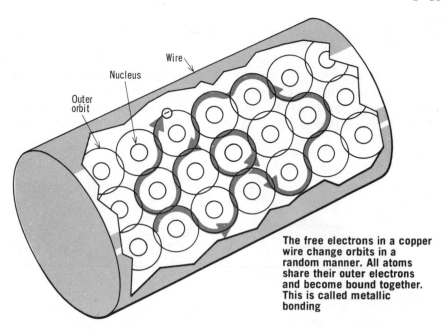

Outer orbit

Nucleus

Wire

The free electrons in a copper wire change orbits in a random manner. All atoms share their outer electrons and become bound together. This is called metallic bonding

free electrons

In order to understand how electrons produce electric current, it would help to review how the atoms of a good conductor, such as copper, are bound together to form the solid metal.

In a copper wire, the atoms each have 1 valence electron, which is loosely held in orbit. And the atoms are close together so that the *outer orbits overlap.* While in motion, the electron of one atom can come under the influence of another atom and enter its orbit. At about the same time, the electron in the second atom is displaced and moves into the orbit of another atom. Most of the outer electrons continually change orbits this way in a *random* manner, so that the valence electrons are not really associated with any one atom. Instead, all of the atoms share all of the valence electrons, and so are bonded together. The electrons are "free" to wander randomly. The action is continuous, so that every atom always has an electron, and vice versa. Therefore, no electric charge results, but the conductor has a lot of free electrons.

In order to produce an electric current, the free electrons in the copper wire must be made to move in the same direction instead of just randomly. This can be done by putting electrical charges on each end of the copper wire; a negative charge at one end, and a positive charge at the other end.

electron movement

Since the electrons are negative, they are repelled by the negative charge and are attracted by the positive charge. Because of this, they cannot change to orbits that would make them move against the forces of the electrical charges. Instead, they *drift* from orbit to orbit toward

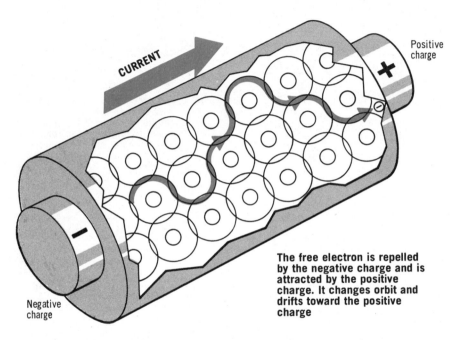

Positive
charge

The free electron is repelled
by the negative charge and is
attracted by the positive
charge. It changes orbit and
drifts toward the positive
charge

Negative
charge

the positive charge, causing an *electric current* to go in that direction.

You can see on the diagram that the density of the atoms in the copper wire is such that the valence orbits of the individual atoms overlap, so that the electrons find it easy to move from one atom to the next. The path that the electron moves in depends on the direction of the orbits that the electron encounters while it drifts to the positive charge. You can see that it does not follow a straight line. But as the charges on each end are made stronger, they control each electron more, causing it to follow a straighter path, and thus move faster through the wire.

The strength of the charge at each end of the wire also determines how many electrons change from a random drifting motion to a more directional drift through the wire. Small charges cause only a small number of electrons to drift to the positive charge. But the stronger the negative and positive charges are made, the more and more electrons will be repelled on a straighter path through the wire.

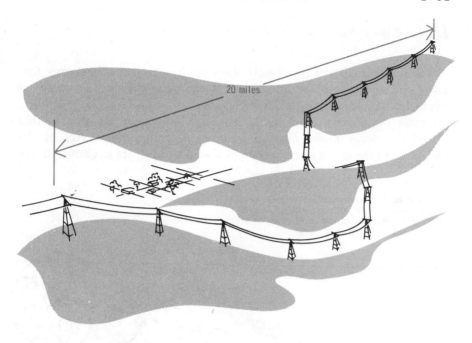

20 miles

If a free electron had to drift down a wire 20 miles long, it could take over 30 days. Yet an electric current travels that distance in a fraction of second

current flow

Although it is sometimes easier to think of the drifting free electrons as being the electric current, it is important to remember this is not completely so. The free *electron* movement *produces* the *current*. This is made clearer when you compare the speed of an electron to the speed of current. The speed of the electron can vary, depending on the conducting material and the number of electrical charges being used. But the speed of the current is always the same.

The free electron that wanders randomly, travels relatively fast because it is only under the influence of the atomic orbital forces; its speed can be a few *hundred miles per second.*

The free electron that drifts under the influence of the electrostatic charges has to oppose some of the atomic orbital forces, and so it is slowed down considerably; it can travel at speeds measured in *inches per second.* This is very slow when you realize that electric current travels at the speed of light: *186,000 miles per second.*

the current impulse

The electric *current* is actually the *impulse* of electrical energy that one electron transmits to another as it changes orbit. When energy is applied to an electron so that it leaves its orbit, it will have to encounter an orbit of another atom as it leaves. The reason for this is that all of the outer orbits overlap and obstruct the free travel of the electron. As the freed electron enters the new orbit, its negative charge

As each electron leaves its orbit and enters another orbit, it repels an electron out of orbit to repeat the action from atom to atom throughout the wire

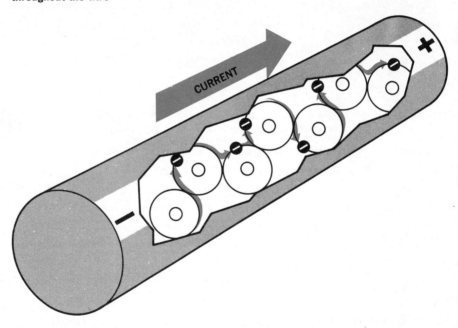

The impulse that is transferred from one electron to the next down the line is the electric current

reacts with the negative charge of the electron already in that orbit. The first electron *repels* the other out of orbit, transmitting its energy to it. The second electron repeats the performance of the first as it encounters the next orbit; and this process continues through the wire. The impulse of energy that is transferred from electron to electron is the electric current.

the speed of electric current

Since the atoms are close, and the orbits overlap, the electron that is freed does not have to travel far to encounter a new orbit. And the moment it enters the new orbit, it transfers its energy to the next electron to free it. The action is almost instantaneous. The same is true for all the following electrons, so that even though each electron is moving relatively slowly, the impulse of electrical energy is transferred down the line of atoms at a very great speed: *186,000 miles per second*. The free electrons are considered *current carriers*.

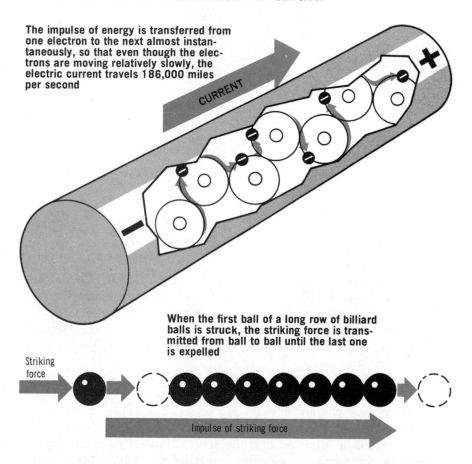

The impulse of energy is transferred from one electron to the next almost instantaneously, so that even though the electrons are moving relatively slowly, the electric current travels 186,000 miles per second

CURRENT

When the first ball of a long row of billiard balls is struck, the striking force is transmitted from ball to ball until the last one is expelled

Striking force

Impulse of striking force

A good analogy of this *impulse transfer* is a long row of billiard balls. When a ball is hit into one end of the row of balls, its striking force is transmitted from ball to ball until the ball at the other end is knocked free. The last ball is released at almost the same time that the first ball is struck.

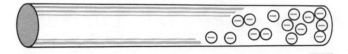

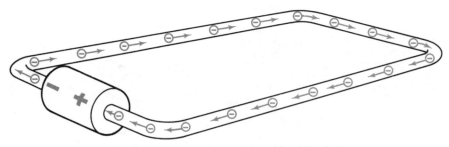

A negative charge put at one end of a wire will repel electrons to the other end until an equal charge is built up to stop electrons from flowing. But, if a power source applies opposite charges to each end of the wire, electrons will continue to flow

An electron leaves the negative side of the battery for each electron that enters the positive side

A COMPLETE OR CLOSED CIRCUIT

a complete (closed) circuit

If a negative charge is put at one end of a wire, that charge would repel free electrons to the other end of the wire. Current would flow only for a moment, until enough electrons accumulate at the other end to produce an equal negative charge that would prevent more electrons from coming. This would be *static electricity* because everything came to rest.

In order to have an electric current, the free electrons must continue to flow; for this to happen, an electrical energy source must be used to apply opposite charges to *each* end of the wire. Then, the negative charge would repel the electrons through the wire. At the positive side, electrons would be attracted into the *source;* but for each electron attracted into the source, an electron would be supplied by the negative side into the wire. Current would, therefore, continue to flow through the wire as long as the energy source continued to apply its electrical charges. This is called a *complete* or *closed circuit.* A battery is a typical electrical source.

A complete or closed circuit is needed for current to flow.

an open circuit

If the wire were broken at any point, electrons would build up at the end of the wire that is connected to the negative side of the battery; and electrons would be attracted away from the end of the wire that is connected to the positive side of the battery. A charge would be built up across the opening to stop the electrons from moving. Current would stop flowing.

An open circuit will not conduct current.

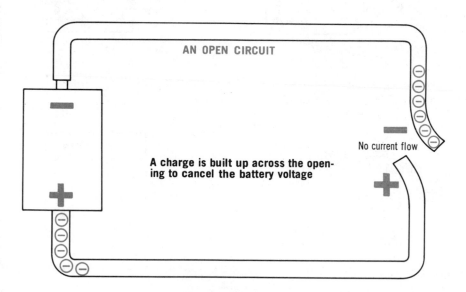

AN OPEN CIRCUIT

No current flow

A charge is built up across the open-ing to cancel the battery voltage

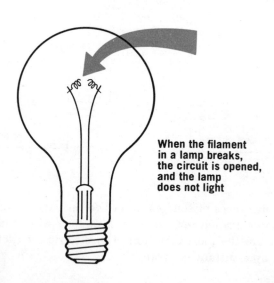

When the filament in a lamp breaks, the circuit is opened, and the lamp does not light

the electrical energy source

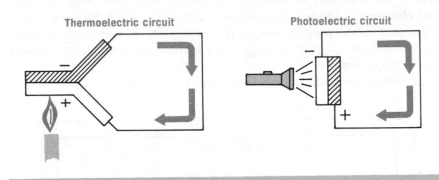

Thermoelectric circuit   Photoelectric circuit

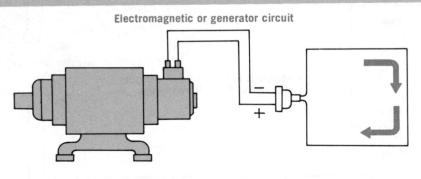

Electromagnetic or generator circuit

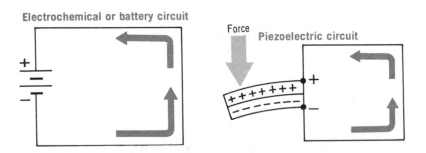

Electrochemical or battery circuit

Piezoelectric circuit

Any one of the five types of sources described on pages 1-42 through 1-46 can be used to cause current to flow through a wire. The battery and the generator are the most common. The electrical outlet in your home is supplied by a distant generator.

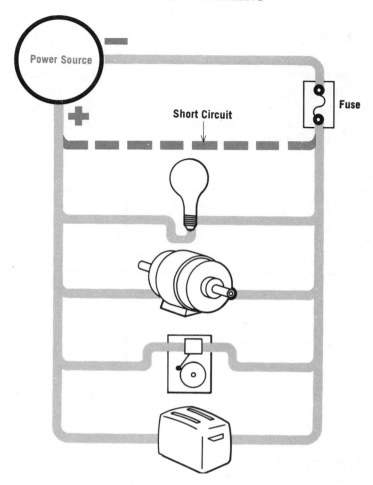

putting electricity to work

Actually, when a good conducting wire is put directly across the *terminals* of a battery or generator, a *short circuit* is produced. Much more current flows than the battery or generator can supply. The battery or generator can burn out and the wire will become very hot. This is why protective fuses are used. They melt when too much current flows, and "open" the circuit.

The wire is used to carry current to other things to make them work. For example, it carries current to heat the filament of a lamp to make it light; it provides electrical energy to turn a motor, ring a bell, heat a toaster, and so on. Some of these uses are explained at the end of this book.

electrical units of measurement

You can see now that two conditions are needed to get current flow: (1) electrical charges to move the free electrons, and (2) a complete circuit to allow electric current to flow. Different amounts of electrical charges can be used, and different amounts of current can flow. There are certain units of measurements that indicate the different values.

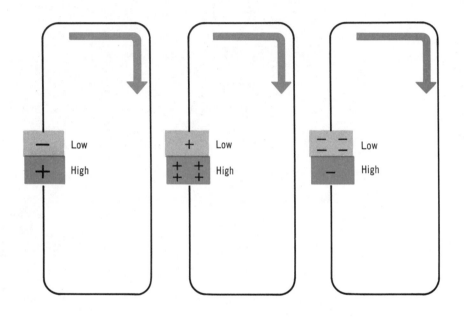

Electron current flows from a low potential to a high potential, or between any difference of potential

The electrical charge that an object gets is called an electric potential, because the electrons that are displaced accumulate potential energy that can be used to move other electrons. Since two charges are needed to complete a circuit, it is the *difference of potential* between these two charges that provides the electric force. *Negative* is considered to be a *low potential,* and *positive,* a *high potential. Electron current* in a wire always goes from the *low to high potential.* This also means that electron current will flow from a low positive potential to a high positive potential, and between two different negative potentials as well.

electromotive force (voltage)

The electrical *charge* that an object gets is determined by the *number of electrons* that the object lost or gained. Because such a vast number of electrons moves, a unit called the *coulomb* is used to indicate the charge. If an object has a negative charge of 1 coulomb, it has gained 6.28×10^{18} (billion billion) extra electrons. This is 6,280,000,000,-000,000,000 electrons.

When two charges have a difference of potential, the electric force that results is called an *electromotive force* (emf). The unit used to indicate the strength of the emf is the *volt*. When a difference of potential causes 1 coulomb of current to do 1 joule of work, the emf is 1 volt. Some typical *voltages* you will probably work with are: 1.5 volts for a flashlight battery, 6 volts for the older auto batteries, 12 volts for the newer auto batteries, 115 volts for the home, 220 volts for industrial power, and so on. Voltages actually vary from *microvolts* (millionths of a volt) to *megavolts* (millions of volts). The terms potential, electromotive force (emf), and voltage are often used interchangeably.

UNITS OF VOLTAGE

A charge of 1 coulomb $= 6.28 \times 10^{18}$ electrons
An emf of 1 volt (v) $=$ 1 coulomb doing 1 joule of work
1 microvolt (μv) $=$ 1/1,000,000 volt
1 millivolt (mv) $=$ 1/1000 volt
1 kilovolt (kv) $=$ 1000 volts
1 megavolt (megav) $=$ 1,000,000 volts

CONVERSION OF UNITS

volts (v) $\times$ 1000 $=$ millivolts (mv)
volts (v) $\times$ 1,000,000 $=$ microvolts (μv)
millivolts (mv) $\times$ 1000 $=$ microvolts (μv)
volts (v) $\div$ 1000 $=$ kilovolts (kv)
volts (v) $\div$ 1,000,000 $=$ megavolts (megav)
megavolts (megav) $\times$ 1000 $=$ kilovolts (kv)

millivolts (mv) $\div$ 1000 $=$ volts (v)
microvolts (μv) $\div$ 1,000,000 $=$ volts (v)
microvolts (μv) $\div$ 1000 $=$ millivolts (mv)
kilovolts (kv) $\times$ 1000 $=$ volts (v)
megavolts $\times$ 1,000,000 $=$ volts (v)
kilovolts $\div$ 1000 $=$ megavolts

1 v $=$ 1000 mv $=$ 1,000,000 μv $=$ 0.001 kv $=$ 0.000001 megav

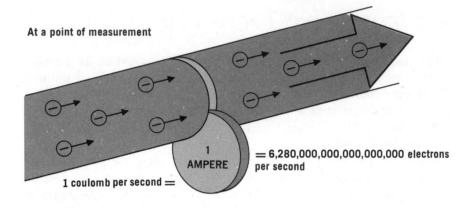

At a point of measurement

1 AMPERE

= 6,280,000,000,000,000,000 electrons per second

1 coulomb per second =

amount of current (*ampere*)

The quantity of current flowing in a wire is determined by the number of electrons that pass a given point in one second. As mentioned previously, a coulomb is 6.28×10^{18} electrons. If 1 *coulomb* passes a point in 1 *second*, then 1 *ampere* of current is flowing. The unit term ampere came from the name of another 18th century scientist, A. M. Ampere. Current is also measured in *microamperes* (millionths of an ampere) and *milliamperes* (thousandths of an ampere).

UNITS OF CURRENT

1 ampere (a) = 1 coulomb/sec
1 milliampere (ma) = 1/1000 ampere
1 microampere (μa) = 1/1,000,000 ampere

CONVERSION OF UNITS

amperes (a) $\times$ 1000 = milliamperes (ma)
amperes (a) $\times$ 1,000,000 = microamperes (μa)
milliamperes (ma) $\times$ 1000 = microamperes (μa)

milliamperes (ma) $\div$ 1000 = amperes (a)
microamperes (μa) $\div$ 1,000,000 = amperes (a)
microamperes (μa) $\div$ 1000 = milliamperes (ma)

0.5 a = 500 ma = 500,000 μa

summary

☐ Metallic bonds are important in the study of electricity. ☐ Electrons in a metal are loosely held in their orbits, and with a little force can be made to enter an overlapping orbit of another atom. ☐ The valence electrons are free to wander from atom to atom. ☐ When a force causes the electrons to move in a specific direction, an electric current is created. ☐ The force which moves the electrons is an electromotive force, emf. It is also called voltage and potential.

☐ The speed of a free electron in random motion can be a few hundred miles per second. Under an emf, its speed will be slowed considerably. ☐ Although the actual speed of an electron under the influence of an emf is slow, the impulse of energy that is transferred from electron to electron is at 186,000 miles per second. This is the rate of current flow. ☐ A current will not flow in a circuit unless there is complete path or loop. Otherwise, the circuit is said to be an open circuit. When a circuit is closed, there is a complete path. ☐ The battery and the generator are the most commonly used sources of force (emf) to transfer electrons. ☐ To prevent an excessive current flow, fuses are used to open the circuit to protect it.

☐ Electron current in a wire always flows from a low to a high potential. ☐ The basic unit of emf is the volt; other units are the microvolt (μv), the millivolt (mv), the kilovolt (kv), and the megavolt (megav). ☐ The amount of 6.28×10^{18} electrons is the coulomb. ☐ The basic unit of electron current is the ampere (a), which is one coulomb per second; other units are the micro-ampere (μa), and the milliampere (ma).

review questions

1. What is electric current, and in what units is it measured?
2. What is a metallic bond?
3. Do orbits of different atoms in a conductor ever overlap?
4. Do electrons under the influence of an emf travel from atom to atom at the rate of 186,000 miles per second?
5. How fast does electron *current* travel? Why is it different from the electron's speed?
6. What is meant by coulomb of charge? Ampere?
7. What is meant by potential difference, voltage, and emf?
8. How many volts are there in 2500 megavolts? 2500 milli-volts?
9. What is a microampere? Milliampere?
10. How does a fuse prevent excessive current from flowing in a circuit?

effects of electricity

Except for friction, electricity can be used to produce the same effects described on page 1-40 that were used to produce electricity.

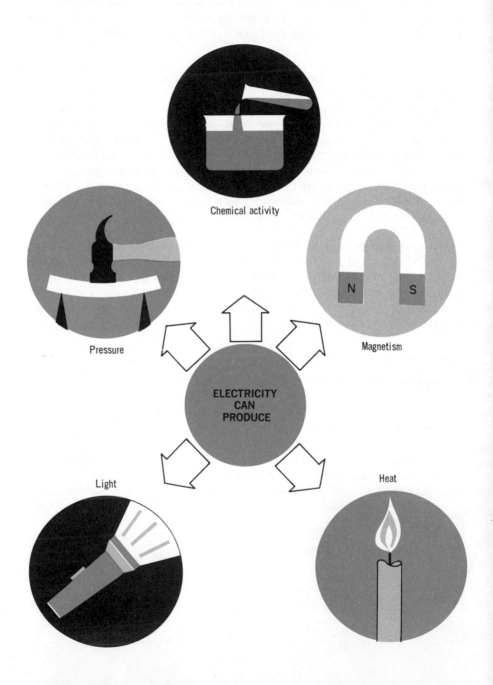

Chemical activity

Pressure

Magnetism

N   S

ELECTRICITY CAN PRODUCE

Light

Heat

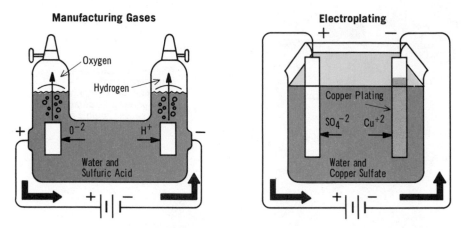

Manufacturing Gases

Oxygen

Hydrogen

O^{-2} H^+

+   −

Water and
Sulfuric Acid

+   −

Electroplating

+   −

Copper Plating

SO_4^{-2} Cu^{+2}

Water and
Copper Sulfate

+   −

The electric potential and current decompose the electrolytes into ions

electricity causes chemical activity

Since the electrical charge is the basic binding force that causes the chemical bonding of compounds, an electric potential or current could be used to alter normal chemical effects. In electrochemistry, this process is called *electrolysis*. For example, if an electric current is passed through water (H_2O) that contains a small amount of sulfuric acid, the water molecules will be broken down into hydrogen and oxygen atoms. The oxygen atoms, though, do not give up the hydrogen atoms' electrons that it previously shared. As a result, the hydrogen atoms become positive ions (H^+), and the oxygen atoms, negative ions (O^{-2}). The ions are attracted to oppositely charged *electrodes*.

At the negative electrode, the positive hydrogen ions pick up electrons, become neutral, and escape the water as a gas. At the positive electrode, the negative oxygen ions give up electrons, become neutral, and escape the water as a gas. The gases can then be bottled. Since an electron comes in at the negative electrode to replace each electron that a hydrogen ion picks up, and one goes out at the positive electrode for each electron that an oxygen atom gives up, current continues to flow until all the water becomes gaseous hydrogen and oxygen.

Another application of electrolysis is *electroplating*. If the water were mixed with copper sulfate ($CuSO_4$), the copper sulfate would break down into positive copper ions (Cu^{+2}) and negative sulfate ions (SO_4^{-2}). The copper ions would go to the negative electrode and pick up electrons. But since copper is a metal, it adheres to the electrode. After a while, the electrode becomes completely plated with copper. Silver and gold plating can also be done this way.

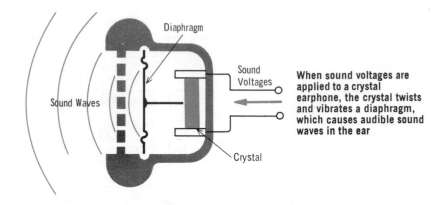

Diaphragm

Sound Voltages

Sound Waves

Crystal

When sound voltages are applied to a crystal earphone, the crystal twists and vibrates a diaphragm, which causes audible sound waves in the ear

electricity causes pressure

Just as a force or pressure that bends or twists some crystals to produce a piezoelectric charge, the application of a voltage will cause the crystal to bend or twist. If an electric potential is placed across a slab of Rochelle salt crystal, the force of the electric field will exert a piezoelectric pressure on the atomic structure and distort the shape of the crystal. This is the way a crystal earphone works, and it is also one method used to "cut" phonograph records.

The crystal cutting head will bend or twist when sound voltages are applied to it. The crystal bends so that the cutting stylus swings the groove to match the sound

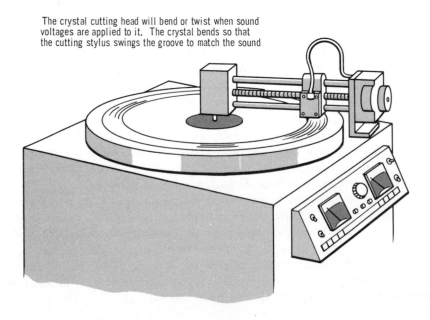

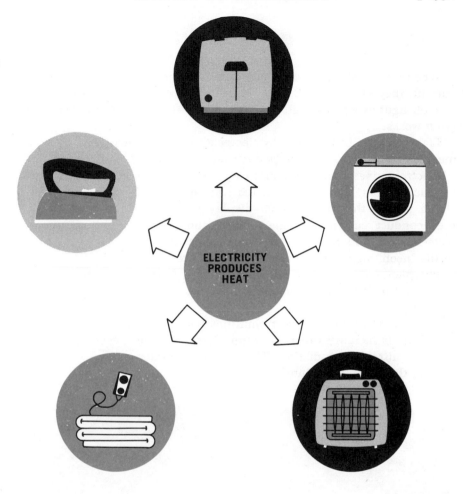

electricity causes heat

Whenever electric current flows through a wire, it produces some *heat*. The reason for this is that some energy is used up in causing the current to flow. This energy is given off in the form of heat. Since it is easiest to cause current flow in good conductors, less heat is produced in good conductors. A poor conductor, such as nichrome, produces a great deal of heat when it conducts current. Copper is about 60 times as good a conductor as nichrome.

The heating effects of electricity are used in many appliances: toasters, irons, electric dryers, electric blankets, heaters, etc.

You should remember, though, that even good conductors produce some heat.

electricity causes light

When many of the poor conductors become hot from conducting current, they glow red and even white hot. Because of this glow, they give off light as well as heat. This is the way the ordinary *incandescent lamp* works.

Light can also be produced by electricity without much heat, by such methods as *fluorescence, phosphorescence,* and *electroluminescence.*

Electroluminescence is produced by some solid materials when they conduct current. The amount of light they give off, though, is relatively slight, and so they are used for display purposes. Many gases, when they conduct current, become ionized and produce light radiations. Neon, argon, and mercury vapor are some examples. They are used in the "neon" signs we see atop our local stores.

Phosphorescence occurs when an electron beam strikes some *phosphors* and other types of materials. The television picture tube works this way.

Fluorescence combines electroluminescence and phosphorescence. A gas, such as mercury vapor, carrying electric current becomes ionized. It emits ultraviolet radiation. The radiation strikes a phosphorescent coating that gives off "white" light.

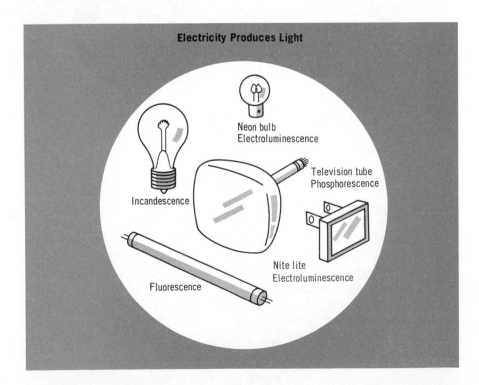

Electricity Produces Light

Incandescence

Neon bulb
Electroluminescence

Television tube
Phosphorescence

Nite lite
Electroluminescence

Fluorescence

Electricity Produces Magnetism

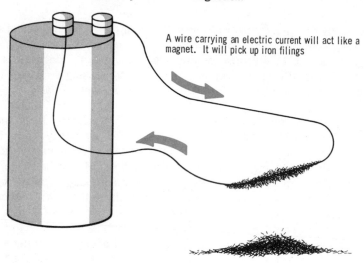

A wire carrying an electric current will act like a magnet. It will pick up iron filings

electricity causes magnetism

Magnetism, too, can be caused by electricity, just as electricity is produced with magnetism. Any conductor that carries an electric current will act like a magnet. This is called *electromagnetism*. Magnetism and electromagnetism are explained more fully in the following pages.

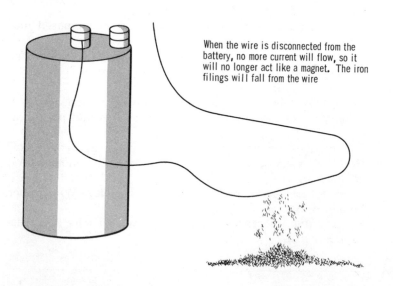

When the wire is disconnected from the battery, no more current will flow, so it will no longer act like a magnet. The iron filings will fall from the wire

summary

☐ Electricity can cause chemical, piezo, thermo, photo, and magnetic effects. ☐ In electrochemistry, the decomposition of chemicals caused by electric current gives rise to electrolysis and electroplating. ☐ Electrolysis is the decomposition of a chemical compound caused by passing a current through its solution. ☐ Electroplating is one application of electrolysis.

☐ If a voltage is applied to certain crystals, a piezoelectric force will exert a pressure that will deform the crystal. ☐ When electricity flows through a poor conductor, it produces heat. ☐ This thermal effect is used in toasters, irons, electric dryers, etc. ☐ When electricity flows, it can be made to produce light. This can be accompanied by considerable heat, as in the common incandescent lamp; or by little heat, as in such methods as fluorescence, phosphorescence, and electroluminescence.

☐ Electroluminescence is produced by gases and some solid materials when they conduct current. "Neon" signs are produced when gases are used for conductors. ☐ Phosphorescence occurs when radiation or an electron beam strikes some phosphors and other types of materials. The television picture tube works on this principle. ☐ Fluorescence combines electroluminescence and phosphorescence. A gas is ionized and emits ultraviolet radiation. The radiation strikes a phosphorescent coating that gives off "white" light. ☐ Any conductor that carries an electric current will act like a magnet. This is electromagnetism.

review questions

1. Name the five effects of electric current.
2. What is electroplating? When a liquid is decomposed into gases by an electric current, what is this called?
3. Which produces more heat: a good or a poor conductor? Name a metal that gives a high thermal effect.
4. What is incandescence?
5. What are electroluminescence and phosphorescence?
6. What process combines electroluminescence and phosphorescence?
7. "Neon" signs are made with _____ as a conductor.
8. How does incandescence differ from the other photoelectric effects?
9. A crystal earphone works on what principle? What other kind of device can use this principle?
10. What is the difference betwen magnetoelectricity and electromagnetism?

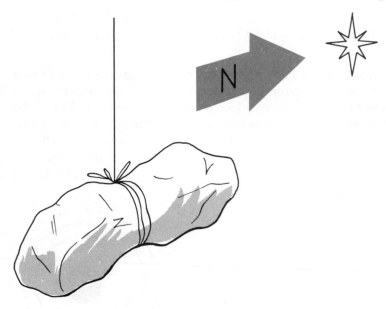

Lodestone was a natural magnet discovered by the
Greeks in Asia Minor over 2000 years ago

magnetism

Magnetism was first discovered over 2000 years ago by the ancient Greeks when they noticed that a certain kind of stone was attracted to iron. Since this stone was first found in Magnesia in Asia Minor, the stone was called *magnetite*. Later, when it was discovered that this stone would align itself north and south when suspended on a string, it was referred to as the *leading stone* or *lodestone*. Lodestone, therefore, is a *natural magnet* that will attract magnetic materials.

magnetism and the electron

Although the forces of electricity and magnetism are related, they are completely different. Magnetic forces and electrostatic forces have no effect on one another as long as there is *no motion*. But, if the force field of either is set in motion, then something happens to cause the two forces to interact. Since the electron is the smallest particle of matter, a theory has been developed to explain the relationship between electricity and magnetism. This is the *electron theory of magnetism*.

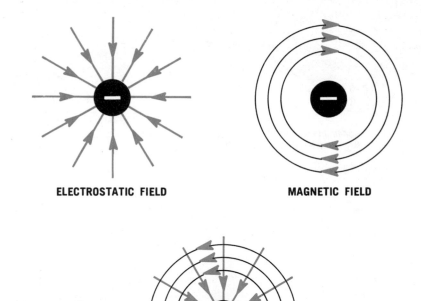

ELECTROSTATIC FIELD   MAGNETIC FIELD

ELECTROMAGNETIC FIELD

We know that the electron contains a negative charge. This charge produces a force field that comes straight in to the electron from all directions. But scientists claim that a *spinning charge* also produces a *magnetic field*. Because of its orbital spin, the electron also contains a magnetic field. But this field exists in concentric circles around the electron. The electrostatic lines of force, then, and the magnetic lines of force are at *right angles* to one another at any one point. The two fields combined are called the *electromagnetic field*.

the magnetic molecule

Actually, iron, nickel, and cobalt are the only naturally magnetic metals; but since all materials contain electrons, you might ask why everything does not have magnetic properties? The answer to this is that the electrons in atoms tend to pair off in orbits with opposite spins, so that their magnetic fields are opposite and cancel each other. But then this might lead one to believe that those elements that have an odd number of electrons are magnetic. If those atoms could be isolated, this could be so; but when atoms combine to form molecules, they ordinarily arrange themselves to produce a total of 8 valence electrons, and in the process, the orbital spins of the electrons cancel the magnetic fields in most materials.

For some reason, though, this orderly process does not occur in iron, nickel, and cobalt. When the atoms of these metals combine, they become ions and share their valence electrons in such a way that many of the electron spins *do not cancel, but add.* This produces regions in the metal called *magnetic domains*, which are referred to as *magnetic molecules*. These magnetic molecules act just like little magnets.

Although iron, nickel, and cobalt are the only naturally magnetic materials, compounds can be manufactured with a controlled process to give them good magnetic properties.

NONMAGNETIC ATOM

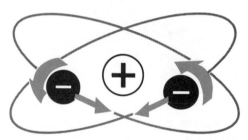

The opposite spins of paired electrons cancel their magnetic effects

In a magnetic molecule, the electrons do not pair off with opposite spins, and the molecule has magnetic properties

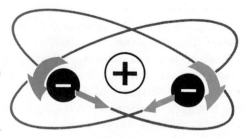

magnetic materials

The naturally magnetic materials are called *ferromagnetic* materials. Ferrous stands for iron, and each of the materials acts like iron with respect to magnetism.

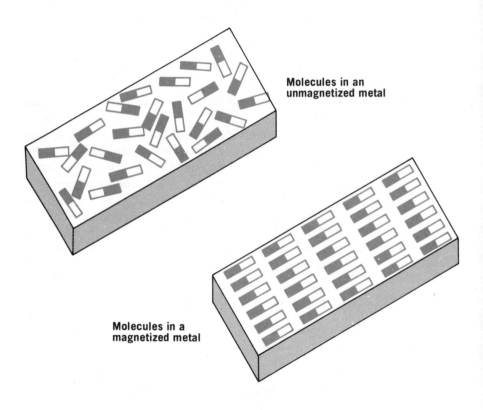

Molecules in an unmagnetized metal

Molecules in a magnetized metal

Since the magnetic materials contain magnetic molecules, it might seem as though they would always act as magnets. But they do not. The reason for this is that under ordinary circumstances, the magnetic molecules are scattered and oriented in a random manner, so that the magnetic fields of the molecules cancel each other. The metal is considered *unmagnetized.*

If all of the molecules were arranged so that they were pointing in the *same direction,* their force fields would *add.* The metal would then be considered *magnetized.* If *all* of the molecules were aligned, a *strong* field would be produced. But, if only *some* of the molecules were aligned, a *weak* magnetic field would be produced. So a magnetic material could also be *partially magnetized.*

how to magnetize iron

Since a magnetic material can be magnetized by *aligning* its molecules, the best way to do it is by applying a magnetic force. The force would act against the magnetic field of each molecule and force it into alignment. This can be done in two ways: (1) by magnetic stroking, and (2) by an electric current.

When a magnet is stroked across the surface of an unmagnetized piece of iron, the field of the magnet aligns the molecules to magnetize the iron.

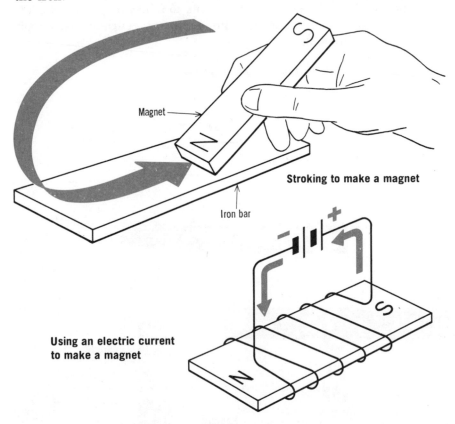

Magnet

Iron bar

Stroking to make a magnet

Using an electric current to make a magnet

When an unmagnetized piece of iron is put in a coil of wire, and the wire is connected across a battery, the electric current produces a magnetic field that magnetizes the iron. This is explained later.

When a magnetized material keeps its magnetic field for a long time, it is called a *permanent magnet*. If it loses its magnetism fast, it is called a *temporary magnet*. Hard iron or steel makes a good permanent magnet. Soft iron is used for temporary magnets.

how to demagnetize a magnet

In order to *demagnetize* a magnet, the molecules must again be *disarranged* so that their fields *oppose* each other.

If the magnet is struck hard, the force would jar the molecules and they would rearrange themselves. Sometimes, repeated strikes are needed.

If the magnet were heated, the *heat energy* would cause the molecules to vibrate sufficiently and rearrange themselves.

If the magnet were placed in a rapidly *reversing magnetic field,* the molecules would become disarranged trying to follow the field. A rapidly reversing field can be produced with alternating current. This topic is covered in Volume 3.

DEMAGNETIZING A MAGNET

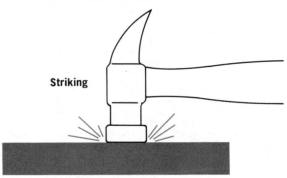

Striking

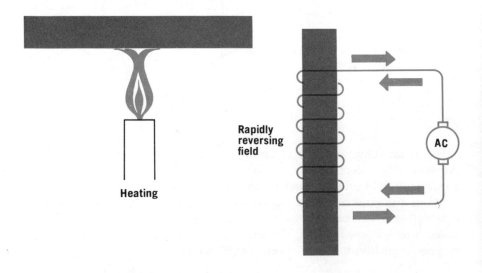

Heating

Rapidly reversing field

AC

the earth's magnetic field

Since the earth itself is a large spinning mass, it too produces a magnetic field. The earth acts as though it has a bar magnet extending through its center, with one end near the north geographic pole, and the other end near the south pole.

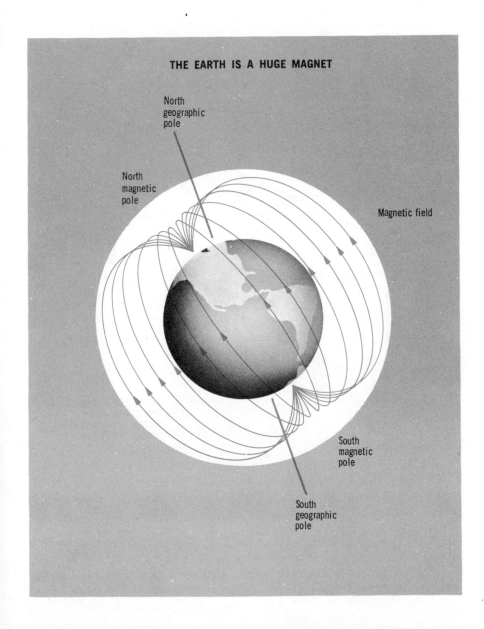

THE EARTH IS A HUGE MAGNET

magnetic polarities

In order to set rules on how magnets affect one another, *polarities* are assigned to the ends of magnets. The polarities are called *north* (N) and *south* (S). The north end of a magnet is determined by hanging that magnet by a string to allow it to swing freely. The magnet will then

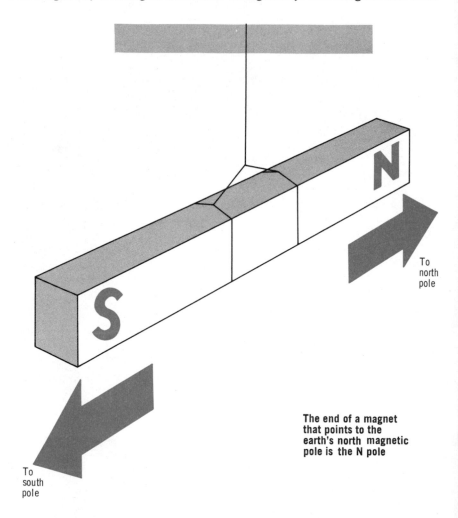

To
north
pole

S

N

To
south
pole

**The end of a magnet
that points to the
earth's north magnetic
pole is the N pole**

align itself with the earth's magnetic field. The end of the magnet that points to the earth's north magnetic pole is called the N pole of the magnet. The other end of the magnet is called the S pole. The magnet will always align itself this way. The reason will be explained later.

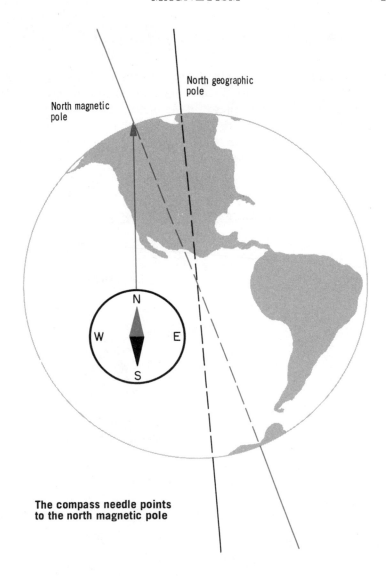

North geographic
pole

North magnetic
pole

**The compass needle points
to the north magnetic pole**

the magnetic compass

Since a magnet will align itself with its N pole pointing north, this
can be used to determine directions. A compass is made with a tiny
light magnet that is freely pivoted so that it will easily keep itself
aligned to the earth's magnetic north pole. Regardless of how the compass is turned, the magnetic needle will point north.

attraction and repulsion

Since a magnet always lines itself up with the earth's magnetic north pole, there seems to be some definite laws governing magnetic effects. These are the laws of attraction and repulsion. The attraction and repulsion laws of magnetism are the same as those of electric charges, except N and S polarities are used instead of negative and positive. The laws are: *Like poles repel, unlike poles attract.*

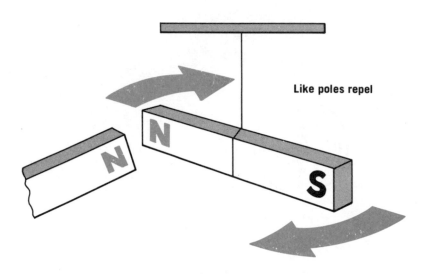

Like poles repel

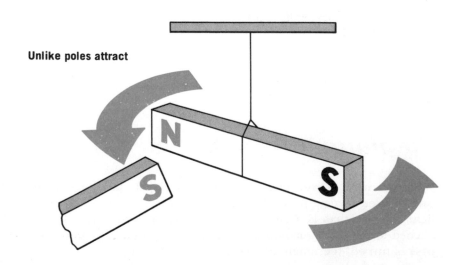

Unlike poles attract

the magnetic field

As you can see by the attraction and repulsion of the magnetic poles, there are forces coming out of the magnetic poles to cause those actions. But the actions do not only take place at the poles. The magnetic *force* actually surrounds the magnet in a *field*. This can be seen when a compass is moved around the bar magnet. In each position around the bar magnet, one end of the compass needle will point to the opposite pole on the bar.

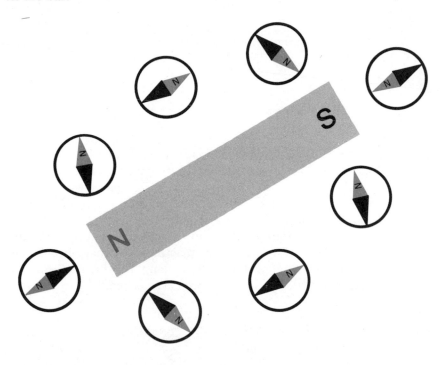

**The compass
shows how the
magnetic force
surrounds a magnet**

The compass can also be used to see how far the magnetic field extends away from the magnet. By withdrawing the compass slowly, you will reach a point where the compass needle is no longer affected by the magnetic field of the magnet, but will again be attracted by the earth's north magnetic pole.

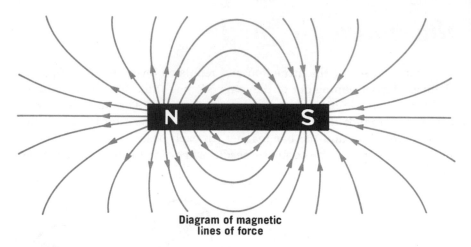

**Diagram of magnetic
lines of force**

lines of force

The magnetic field of a magnet is made up of *lines of force* that extend out into space from the N pole of the magnet to the S pole. These lines of force do not cross, and they become wider apart away from the magnet. The closer the lines of force are and the greater the number of force lines, the stronger the magnetic field.

The existence of the lines of force can be demonstrated by sprinkling iron filings on a flat surface, and then placing a bar magnet on them. The iron filings will arrange themselves along the lines of force to show the magnetic field. The lines of force are also called *flux lines*.

**Fine iron filings show up the
lines of force of a bar magnet**

interaction of magnetic fields

When two magnets are brought together, their fields interact. Magnetic lines of force *will not cross* one another. This fact determines how fields act together.

If the lines of force are going in the *same direction*, they will *attract* each other and join together as they approach each other. This is why unlike poles attract.

If the lines of force are going in *opposite directions*, they *cannot combine*. And, since they cannot cross, they apply a force against each other. This is why like poles repel.

The interaction of the flux lines can also be shown with iron filings.

Unlike poles attract

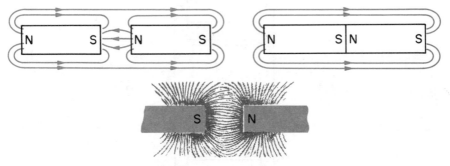

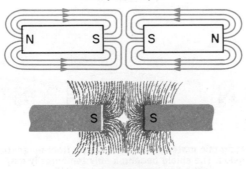

Like poles repel

**Iron filings show
the flux lines**

magnetic shielding

Magnetic flux lines can pass through all materials, even those that have no magnetic properties. Some materials, though, resist the passage of flux lines somewhat. This property is known as *reluctance*. Magnetic materials have a very low reluctance to flux lines. The lines of flux will be attracted through a magnetic material even if it has to take a longer path. This characteristic allows us to shield things from magnetic lines of force by enclosing them with a magnetic material. This is the way antimagnetic watches are made.

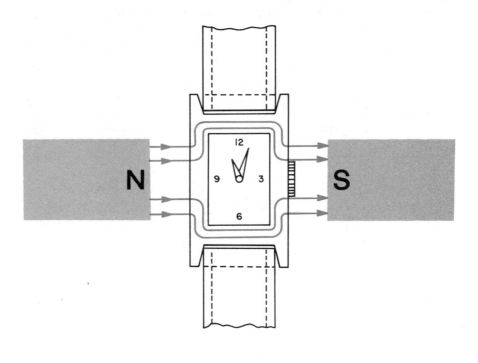

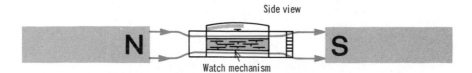

Side view

Watch mechanism

Magnetic materials shield a watch from magnetic fields. The shield becomes only temporarily magnetized. The flux lines cannot reach this watch mechanism

summary

☐ The interaction of electricity and magnetism to form an electromagnetic field is explained by the electron theory of magnetism. ☐ Atoms of certain metals combine so that their valence electrons are shared in such a way to form magnetic domains or molecules. These are examples of magnetic materials. ☐ A material with its magnetic domains or molecules aligned is said to be magnetized. ☐ A magnetic material can be magnetized by applying a magnetic force either by stroking or by an electric current. ☐ A magnetic material can be demagnetized by being heated, struck, or by being placed in a rapidly reversing magnetic field.

☐ The earth produces a magnetic field. ☐ North (N) and south (S) polarities are assigned to magnets. ☐ The N pole of a freely pivoted magnet will point to the earth's north pole. The other pole is the S pole. A compass utilizes this principle. ☐ The laws of repulsion and attraction for magnets are: like poles repel, and unlike poles attract.

☐ Magnetic force surrounds the magnet in a magnetic field. ☐ The magnetic field is made up of lines of force that extend into space from the N pole of the magnet to the S pole. ☐ Lines of force are called flux lines. ☐ Lines of force do not cross. ☐ The closer the lines of force and the greater the number of force lines, the stronger the magnetic field. ☐ Lines of force in the same direction attract and join; this is why unlike poles attract. ☐ Lines of force in opposite directions repel and cannot combine; this is why like poles repel.

review questions

1. What are magnetic domains or molecules?
2. How can a magnet be demagnetized?
3. What is a ferromagnetic material?
4. What is meant by the reluctance of a material? Why should the reluctance of a shield be smaller than the shielded material?
5. What is an electromagnetic field?
6. How can a ferromagnetic metal be magnetized?
7. What are the laws of repulsion and attraction for magnetic poles?
8. What are flux lines?
9. Which pole of a compass needle points to the earth's north pole?
10. What is the difference between permanent and temporary magnets?

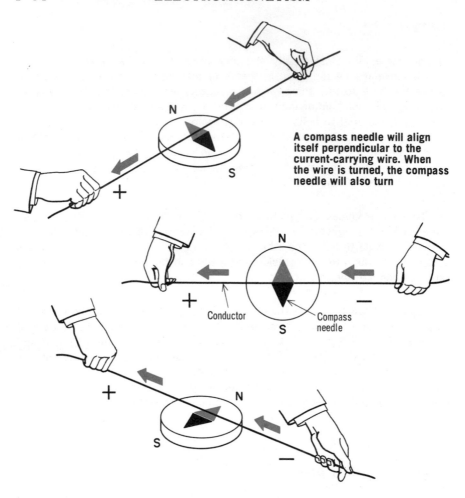

A compass needle will align itself perpendicular to the current-carrying wire. When the wire is turned, the compass needle will also turn

what is electromagnetism?

Since the electron produces its own magnetic field because of its orbital spin, the accumulation of surplus electrons in an object might seem to be able to produce a magnetic field. But, again, with static charges, electrons with opposite spins pair off to cancel their magnetic effects. Static electricity, then, has no magnetic field.

Electrons moving through a wire under a force causing a *current flow*, though, cannot pair off with opposite spins. On the contrary, since they are drifting in the same direction, their magnetic fields tend to add.

In 1819, Hans Christian Oersted discovered that an electric current produced a magnetic field when he noticed how a wire carrying a current affected a compass.

electromagnetism in a wire

Since the magnetic field around an electron forms a loop, the fields of the electrons combine to form a series of loops around the wire. The direction of the magnetic field depends on the direction that the current flows. A compass moved around the wire will align itself with the flux lines.

A *left-hand rule* can be used to determine the direction of the magnetic field. If you wrap your fingers around the wire with your thumb pointing in the direction of electron current flow, your fingers will point in the direction of the magnetic field.

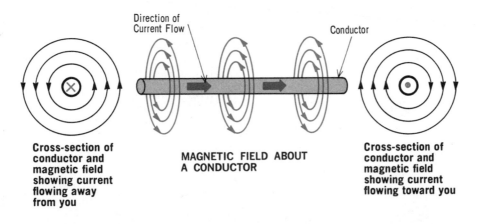

Direction of Current Flow   Conductor

Cross-section of conductor and magnetic field showing current flowing away from you

MAGNETIC FIELD ABOUT A CONDUCTOR

Cross-section of conductor and magnetic field showing current flowing toward you

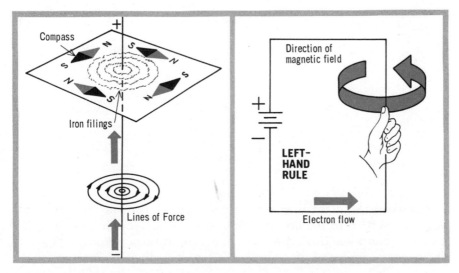

Compass

Iron filings

Lines of Force

Direction of magnetic field

LEFT-HAND RULE

Electron flow

field intensity

The more current that flows through a wire, the stronger the magnetic field will be. As with the magnet's magnetic field, the flux lines are closer together near the wire, and they move further apart as they move away from the wire. The field, then, is stronger near the wire, and becomes weaker with distance.

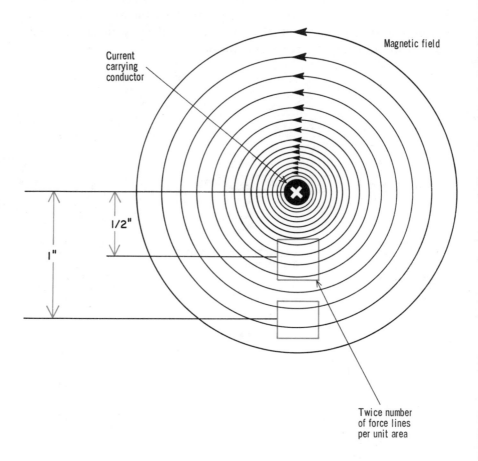

The decrease in the number of lines of force per unit area is in inverse proportion to the distance from the conductor. That is, at a distance of one inch from the conductor, for instance, there is one-half the density of force as at a distance of one-half inch.

field interaction

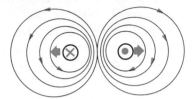

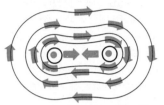

**Opposite currents cause
opposite fields that repel**

**Currents in the same direction
cause fields to add and attract**

If two wires that are carrying current in opposite directions are brought close together, their magnetic fields will oppose one another, since the flux lines are going in opposite directions. The flux lines cannot cross, and the fields move the wires apart.

When wires that are carrying current in the same direction are brought together, their magnetic fields will aid one another, since the flux lines are going in the same direction. The flux lines join and form loops around both wires, and the fields bring the wires together. The flux lines of both wires add to make a stronger magnetic field. Three or four wires put together in this way would make a still stronger field.

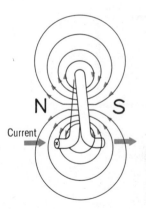

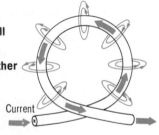

When a wire is being formed into a loop, all of the circular fields enter one side of the loop and leave the other side

The flux lines are compressed in the center of the loop to create a strong field. The N pole is produced on the side that the flux lines come out

If the wire is twisted to form a *loop*, the magnetic fields around the wire will all be arranged so that they each flow into the loop on one side, and all come out on the other side. In the center of the loop, the flux lines are compressed to create a dense and stronger field. This produces magnetic poles, with north on the side that the flux lines come out, and south on the side that they go in.

electromagnetism in a coil

If a number of loops are wound in the same direction to form a *coil*, more fields will add to make the flux lines through the coil even more dense. The magnetic field through the coil becomes even stronger. The more loops there are, the stronger the magnetic field becomes. If the coil is compressed tightly, the fields would join still more to produce a still stronger *electromagnet*.

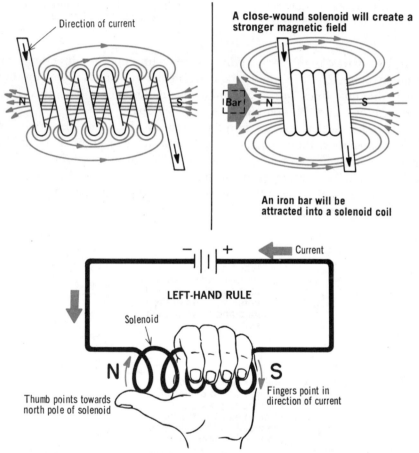

Direction of current

A close-wound solenoid will create a stronger magnetic field

N   S

Bar   N   S

An iron bar will be attracted into a solenoid coil

— | | | +   Current

LEFT-HAND RULE

Solenoid

N   S

Thumb points towards north pole of solenoid

Fingers point in direction of current

A helically wound coil that is made to produce a strong magnetic field is call a *solenoid*. The flux lines in a solenoid act the same as in a magnet. They leave the N side and go around to the S pole. When a solenoid attracts an iron bar, it will draw the bar *inside* the coil.

There is a *left-hand rule* for solenoids too. If you wrap your fingers around the coils in the direction of electron current flow, your thumb will point to the N pole.

the magnetic core

The magnetic field of a coil can be made stronger still by keeping an *iron core* inside the coil of wire. Since the soft iron is magnetic and has a low *reluctance,* it allows more flux lines to be concentrated in it than would air. The more flux lines there are, the stronger the magnetic field. Soft iron is used as a core in an electromagnet because hard iron would become permanently magnetized.

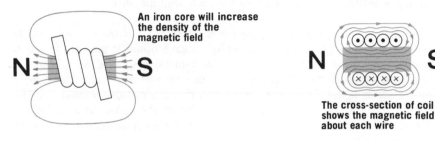

An iron core will increase the density of the magnetic field

The cross-section of coil shows the magnetic field about each wire

magnetomotive force

The magnetizing force that is caused by current flowing in a wire is called the *magnetomotive force* (mmf). The mmf depends on the current flowing in the coil and the number of turns in the coil. If the current is doubled, the mmf will be doubled. Also, if the number of turns (loops) in the coil is increased, the mmf will be increased.

The mmf, then, is determined by a term called ampere-turns, which is the electric current multiplied by the number of turns of the coil.

The magnitude of the mmf determines the number of lines of flux there will be in the field, or how strong the field will be. As the mmf is increased, the number of flux lines also increases. But there will be a point at which the mmf being increased will no longer produce more flux lines. This is known as the *saturation point.*

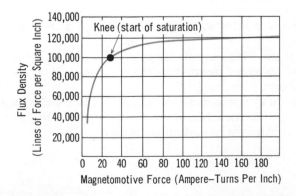

summary

☐ Electrons produce magnetic as well as electrostatic fields. But normally, electrons cancel each other's magnetic effects. ☐ Electrons moving through a wire because of current flow produce a magnetic field because the electrons' fields aid each other. ☐ The direction of the magnetic field depends on the direction of the current flow, according to the left-hand rule: If your fingers are wrapped around the wire, with your thumb in the direction of the electron current flow, they will indicate the direction of the magnetic field. ☐ The stronger the electron current, the stronger is the magnetic field. ☐ The field is stronger near the wire, and becomes weaker with distance.

☐ The fields of two current-carrying wires interact. ☐ If the currents are in the same direction, the wires are attracted to each other; if they are in opposite directions, they repel. ☐ A loop of current-carrying wire has a concentrated field at its center. ☐ Many loops can be helically wound to form a strong electromagnet. This electromagnet is called a coil or solenoid. ☐ The left-hand rule for a solenoid gives the direction of the poles: If your fingers are wrapped around the wire in the direction of the electron current flow, your thumb will point to the N pole.

☐ Soft iron cores are used to concentrate and strengthen the lines of flux. ☐ The magnetizing force that is caused by current flowing in a wire is called the magnetomotive force, mmf. ☐ Ampere-turns is the unit of mmf. It is equal to the number of loops or turns of wire multiplied by the current in amperes. ☐ When an increasing mmf no longer increases the number of flux lines, the coil is saturated.

review questions

1. State the rule to determine the direction of the magnetic field around a current-carrying conductor.
2. How does the current magnitude affect the field magnitude?
3. How does the number of lines of force vary with distance?
4. What happens when two current-carrying wires are brought together?
5. What is a coil or solenoid? What does it do for the magnetic field?
6. State the left-hand rule for solenoids.
7. Why are cores inserted in electromagnets? Are steel cores ever used?
8. How do electromagnets act compared to regular magnets?
9. What is magnetomotive force, and how is it measured?
10. What is meant by saturation point?

putting electricity
and magnetism to work

Because of the various effects that electricity has, and the way it is related to magnetism, these two types of energy can be applied in many ways to do work.

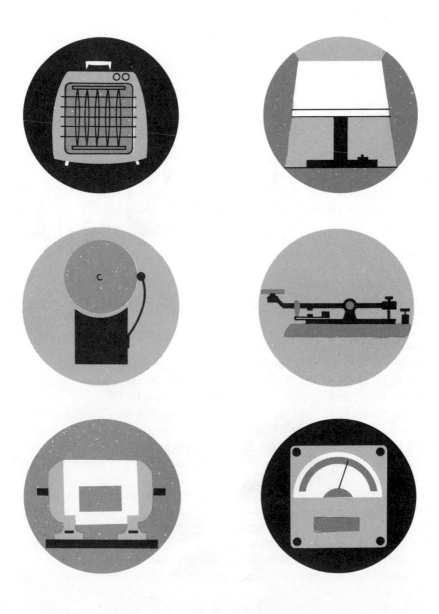

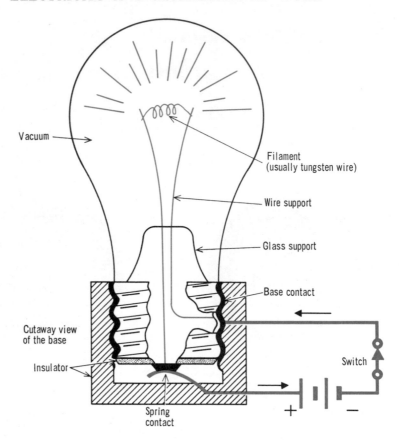

Vacuum

Filament
(usually tungsten wire)

Wire support

Glass support

Base contact

Cutaway view
of the base

Insulator

Switch

Spring
contact

+   −

the incandescent light

As explained earlier, the incandescent lamp works because electricity that flows through a poor conductor causes it to glow red or white hot; and light is given off. That poor conductor in the light bulb is called the *filament,* and is usually made of tungsten wire.

The filament is held up by two wire supports that go down through a piece of glass support to the base. One wire goes to the brass threads, and the other wire goes to a metal button at the bottom. The threaded brass and the button are good conductors and are separated by an insulator.

When the switch is closed, the circuit is complete. Electron current flows from the negative side of the battery, through the switch and base contact, and up one support wire. The current then goes through the filament, down the other wire support to the button; and then through the spring contact to the positive side of the battery.

The air is taken out of the bulb to keep the filament from burning up.

the electric heater

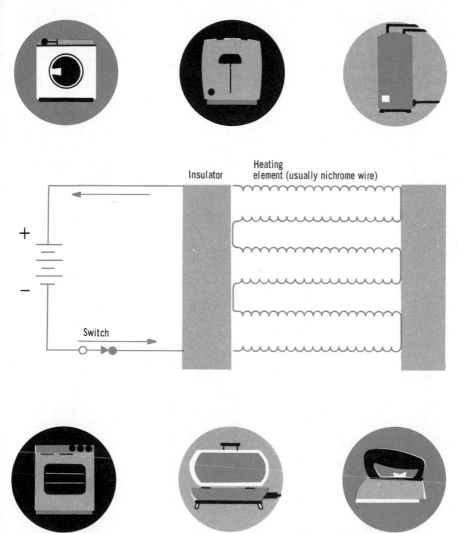

The electric heater is made similar to the incandescent light bulb, except that the material used for the heating element may not glow too brightly. When the switch is closed, the current from the battery goes through the heating element to heat it. It usually gives off a soft red glow, and heats the surrounding air. The heating element, which is usually nichrome wire, is supported by insulation.

Electric clothes dryers, hot water heaters, stoves, irons, toasters, grills, etc. work in similar ways.

the electromagnetic relay

The *electromagnetic relay* uses the action of a magnetic field to attract a movable contact against a fixed contact to close another circuit.

When the switch is closed, the battery sends electron current through a coil that is wrapped around a soft iron core. The current in the coil produces a magnetic field that magnetizes the core. The magnetic field from the core attracts the movable contact, which is also made of magnetic material, and pulls it down firmly against the fixed contact. The two contacts close just like a switch to energize the other circuit. When the main switch is opened, the current stops going through the coil, and the magnetic field collapses. The movable contact is released, and springs away from the fixed contact to open the other circuit.

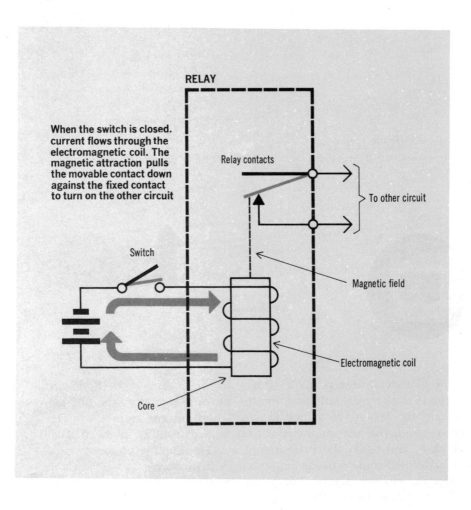

RELAY

When the switch is closed. current flows through the electromagnetic coil. The magnetic attraction pulls the movable contact down against the fixed contact to turn on the other circuit

Relay contacts

To other circuit

Switch

Magnetic field

Electromagnetic coil

Core

the electromagnetic relay (cont.)

Sometimes it is desirable to want the relay to stay energized when the operating switch is released. A special relay, called a *holding relay*, is used, which has an extra set of contacts to keep current flowing

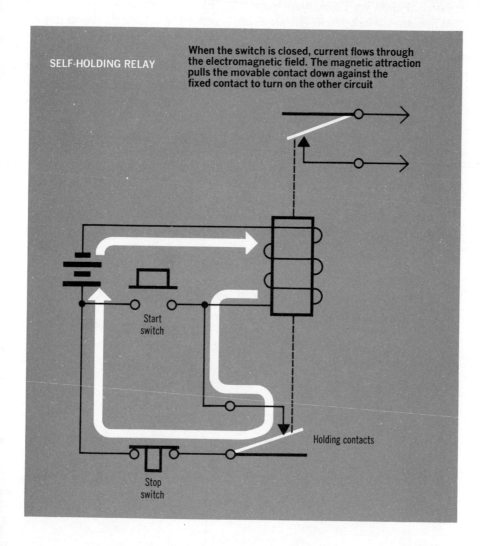

SELF-HOLDING RELAY

When the switch is closed, current flows through the electromagnetic field. The magnetic attraction pulls the movable contact down against the fixed contact to turn on the other circuit

Start switch

Stop switch

Holding contacts

through the relay even after the main switch is released. With this kind of relay, another switch must be used to shut the relay off. It is called a *stop switch*, and when it is pushed it opens the circuit to the *holding contacts*.

the buzzer

The *buzzer* operates similarly to the relay, except that the contacts are specially arranged so that the *electromagnetic coil* cannot stay energized continuously. A movable armature which normally stays pressed against an adjustment screw forms the relay contacts.

When the switch is closed, electron current goes through the contact screw and the armature to energize the coil. The coil produces a magnetic field that attracts the armature away from the screw. When that happens, the circuit is opened, the magnetic field collapses, and the springy armature swings back against the screw. As soon as the armature touches the screw again, the circuit is closed, current flows again, and the magnetic field pulls the armature away again to open the circuit.

So you can see that the armature moving up and down, on and off the screw, opens and closes the circuit continuously.

The armature chatters against the screw, producing a buzzing sound. By adjusting the screw, you can control how far and how fast the armature vibrates, thus adjusting the kind of buzzing sound you will get.

A *bell* works similarly to the buzzer. The bell uses a long swinging hammer attached to the armature, so that the hammer can be vibrated against a bell.

The armature and adjustment screw act as a switch that opens and closes every time the armature is attracted to the coil. The armature vibrates up and down against the screw to produce a buzzing sound

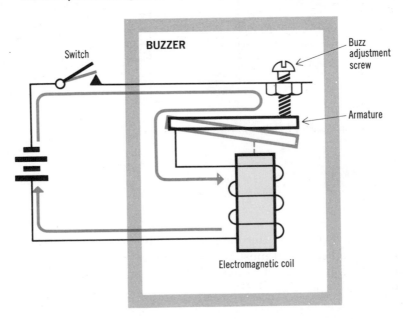

Switch

BUZZER

Buzz adjustment screw

Armature

Electromagnetic coil

the basic telegraph set

The diagram below shows how a basic telegraph set works. The telegraph *keys* are used as *switches* that open and close the circuit to form dots and dashes.

If the "east" station wants to receive from the "west" station, he must hold his key down. Then, when the west station puts his key down, current will flow through the relays and energize them. The relays will close their armature contacts to connect the sounder batteries to the sounder relays. The sounder relay armatures vibrate when they are energized just like the buzzer described previously. If the west station's key is held down for a *short* time, he sends a *dot*. But, if the key is held down *longer,* he sends a *dash*. Both sounder relays work together, so the west station can also hear what he *transmits*.

When the east station wants to send a message, he opens his key to open the entire circuit. Then, when the west station hears his sounder stop, he knows that the east station wants to send; the west station then closes his key and listens.

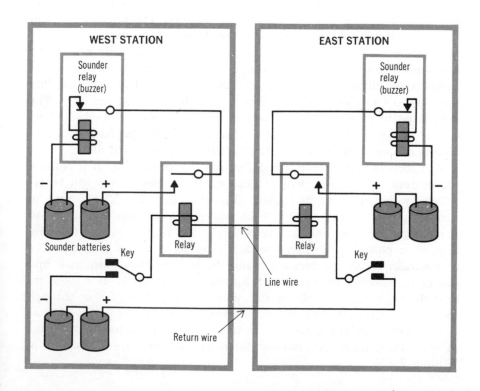

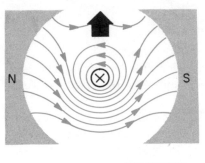

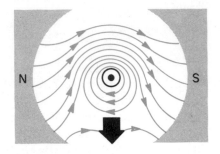

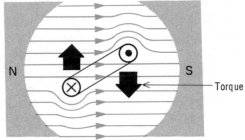

the electric motor

The electric motor works because of the effect that a magnetic field has against a wire carrying an electric current. The current through the wire produces its own magnetic field around the wire. This field will distort the flux lines that exist between two magnetic poles. The flux lines will tend to move to the side of the wire where they are going in the same direction as the wire's lines of force. The distorted flux lines try to straighten, and so exert a repelling force on the wire. The wire is pushed out of the field where the flux lines are weakest. This is the principle of electric motor operation. The *right-hand rule* for motors can be used to find the direction that the wire will move.

If a loop of wire is connected through a *commutator* to a battery, the current in that wire will produce magnetic fields that will be re-pelled by the magnet's flux lines. This will cause the looped wire to turn, or produce a *torque*. When the loop gets to the position shown in B on page 1-99, the repelling force stops, but inertia carries the wire around to position C, where the field's repulsion turns it again. The commutator is needed because, when the loop reaches position B and passes it, it would be repelled back to position A again. But the com-mutator segments are split at that point, so that the current through the wire is reversed, and the wire is repelled in the same direction as before. The rotor (or armature) in this type of motor has many turns of wire and many commutator segments. Motors are examined in Volume 7.

the electric motor (cont.)

RIGHT-HAND RULE

To determine the direction in which a current-carrying wire will move in a magnetic field, use the right hand motor rule. Point the index-finger in the direction of the magnetic field and the second finger in the direction of current flow. The thumb will indicate the direction of motion

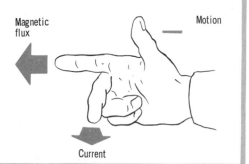

Magnetic flux

Motion

Current

(A)

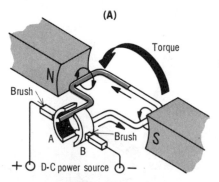

Torque

Brush

N

A

Brush

B

S

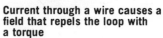
+ ○ D-C power source ○ −

Current through a wire causes a field that repels the loop with a torque

(B)

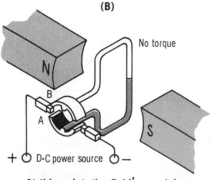

No torque

N

B

A

S

+ ○ D-C power source ○ −

At this point, the field's repulsion stops, but inertia carries the loop toward diagram (C)

(C)

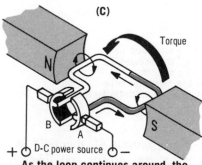

Torque

N

B

A

S

+ ○ D-C power source ○ −

As the loop continues around, the commutator reverses the current direction so that the magnetic field repels the wire in the same direction

(D)

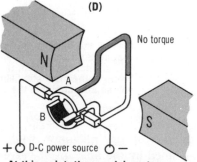

No torque

N

A

B

S

+ ○ D-C power source ○ −

At this point, the repulsion stops again, but inertia carries the loop toward position A to repeat the cycle

the meter

A basic type of meter can use a solenoid coil and a movable core to measure the current that is flowing. Whenever current flows through the coil, a magnetic field is built up, which attracts the core into it. However, the other end of the core is pivoted with a spring that tends to hold it back. The distance that the core moves, depends on the

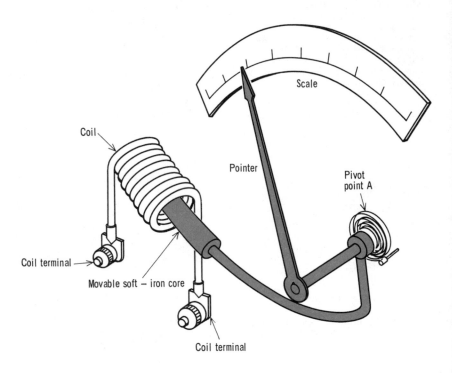

Current through the coil produces a magnetic field that attracts the core in and swings the pointer. The greater the current through the coil, the stronger the magnetic field, and the more the core will be attracted into the coil to swing the pointer further

strength of the magnetic field; and the strength of the field is determined by the amount of current flowing through the coil. Therefore, the core moves farther into the coil when more current flows.

The pivot of the core also contains a pointer that swings along a scale to indicate the current that is being measured. Meters are examined in Volume 5.

the basic generator

The generator's operation is opposite to that of the motor. Instead of putting a current in the rotor windings to produce a magnetic field, the rotor is *turned mechanically*, usually with a motor.

Then when the rotor windings pass through the lines of flux, the magnetic energy forces current to flow in the wire. When the wire goes down the field, current flows in one direction; but when the wire goes up the field, the current flows in the other direction. The commutator, though, switches the wires outside the generator while the rotor turns, to keep the current flow in the *same direction* through the meter at all times. Therefore, it is called a *direct-current* (d-c) generator. If a commutator is not used, the current coming out of the generator will change direction as the loop turns. This is called *alternating current* (ac). It is examined in Volume 3.

Current is induced in a wire that is moved through a magnetic field. The direction of the current flow depends on the direction the wire is moved

The commutator keeps the current flowing out of the generator in the same direction at all times

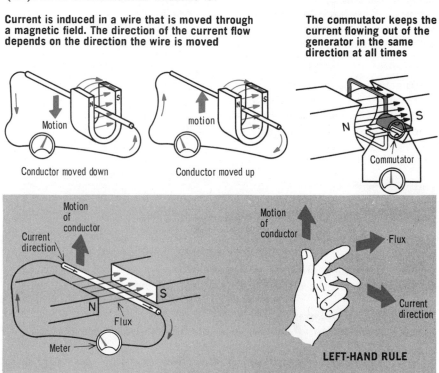

Conductor moved down

Conductor moved up

Commutator

LEFT-HAND RULE

The *left-hand rule* for generators can be used to determine the direction of current flow that will result when a wire is moved through a magnetic field.

Generators are examined in detail in Volume 6.

summary

☐ Some electrical energy applications are: the incandescent lamp and the electric heater. ☐ The incandescent lamp gives off light when current flows through a poor conductor. The conductor or filament, is usually made of tungsten. ☐ The electric heater uses the same principle as the incandescent lamp, but the heating element does not glow as brightly. The heating element is usually nichrome. ☐ Other devices using heating elements are irons, toasters, grills, etc.

☐ Some magnetic energy applications are: the electromagnetic bell, the relay, the basic telegraph set, and the electric motor. ☐ The electromagnetic bell uses the action of a magnetic field to vibrate an armature so that a hammer strikes a bell. ☐ A relay is an electromagnet that opens or closes contacts. ☐ A basic telegraph set consists of keys that act as switches to open and close the circuit by relay action to form dots and dashes. ☐ The electric motor operates because of the effect of a magnetic field on a current-carrying wire. The interaction causes the rotation of an armature. The right-hand motor rule gives the direction that the rotor loop will rotate.

☐ The basic meter is an instrument that uses a solenoid coil to attract a movable core to measure current. ☐ The basic generator produces electricity. Its operation is opposite to that of the motor. The left-hand generator rule gives the direction of the current generated.

review questions

1. How does an incandescent lamp differ from an electric heater?
2. State the right-hand motor rule.
3. What do the electromagnetic bell and buzzer have in common?
4. Why is a commutator used in certain motors?
5. Is a commutator always necessary for a generator to work?
6. State the left-hand generator rule.
7. How can a meter measure current?
8. What is the difference between direct current and alternating current?
9. Describe a basic telegraph set.
10. What is the difference between a magnetic circuit breaker and a fuse?

index